scholarly journals
IN THE NEW
DIGITAL
WORLD

scholarly journals
IN THE NEW
DIGITAL
WORLD

Gérard Boismenu
Guylaine Beaudry

translated from French by
Maureen Ranson

UNIVERSITY OF
CALGARY
PRESS

Original title: Le nouveau monde numérique

Copyright © 2002 by Les Presses de l'Université de Montréal, Montréal, Québec, Canada

English translation: Copyright © 2004 by Maureen Ranson
Published by the University of Calgary Press
2500 University Drive NW, Calgary,
Alberta, Canada T2N 1N4
www.uofcpress.com

We acknowledge the financial support of the Government of Canada through the Book Publishing Industry Development Program (BPIDP) for our publishing activities. We acknowledge the support of the Alberta Foundation for the Arts for this published work.

This book has been published with the help of a grant from the Canadian Federation for the Humanities and Social Sciences, through the Aid to Scholarly Publications Programme, using funds provided by the Social Sciences and Humanities Research Council of Canada.

Canada

**Canada Council   Conseil des Arts
for the Arts   du Canada**

National Library of Canada Cataloguing in Publication

Boismenu, Gérard
    Scholarly journals in the new electronic world / Gérard Boismenu,
Guylaine Beaudry.

Translation of: Le nouveau monde numérique : le cas des revues
    universitaires.
Includes bibliographical references.
ISBN 1-55238-127-7

    1. Electronic journals--Publishing. 2. Electronic journals--Publishing--Canada. 3.
Scholarly periodicals--Publishing. 4. Scholarly periodicals--Publishing--Canada. 5.
Scholarly electronic publishing. 6. Scholarly electronic publishing--Canada. I. Beaudry,
Guylaine II. Title.

PN4833.B6413 2004   070.5'72'0285   C2004-901483-8

Printed and bound in Canada by Houghton Boston
∞ This book is printed on 50% recycled, acid-free paper

Cover and page design by Mieka West, typesetting by Elizabeth Gusnoski

To our children François Frédéric, Philippe and Marie

CONTENTS

CONTENTS

FOREWORD

The English version[1] of this book will reach a new audience and allow participation, however modest, in reflection on the transformation of scientific communication brought about by digitization. The book's focus is on scholarly journals, their place and contribution to renewed forms for the use, production, dissemination and preservation of research results.

Why we should deal with this issue might be questioned, if, in the minds of many, it has been or should have been settled a long time ago. We have only to observe the situation in Canada and in many European countries to realize that "digital time" or even "prophetic or utopian time" passes at a faster pace than the "real time" needed for social transformation in modes of expression and communication in a community that has its own rules, practices and institutional environment.

Social transformation is not a simple act of will, but rather a lengthy process that calls a diverse set of factors and participants into play. Otherwise, how can we explain the fact that the transition of journals to the digital age has not yet been accomplished by many national infrastructures of scientific communication?

Canada is a good example, yet many European countries could be cited as well. Equipped with world-class national research structures and with researchers who operate on a global level, these countries have still not succeeded in providing journals that are nationally rooted organs of knowledge dissemination with the means to assume their place in the digital world. This is neither fate nor the work of decision makers seeking to keep them in a knowledge hinterland. The reasons must be sought elsewhere. Moreover, criteria such as the contribution and relative rank of national journals in scientific communication also fail to provide answers, as they are subject to tireless solicitation by the large commercial enterprises trying to draw them into their bundle of services and databases.

It would also be too easy to lay blame on a single category of players or a single factor. Sociological fiction of that kind is neither fruitful nor useful in understanding what is happening. In this book, we aim to survey the various players and the dynamics of the constellation and aspects of information science that increase overall comprehension. This approach is never finished, always a work-in-progress. By definition, we subscribe to a concept in movement. Things change. Even the way we perceive change changes. With a little distance and an awareness of how long it takes for initiatives to be organized, established and confirmed, whether in Canada, Europe or the European Union, as institutions or community policies, we would like to highlight a component here we did not want to over-emphasize before — the involvement of public authorities, a dimension that demonstrates the social and political aspect at work in advancing initiatives broad in scope and influence.

In view of the large commercial groups that charge exorbitant prices and pocket disproportionate profit margins, responsible publishers and non-profit organizations must find the resources to sustain an alternative that is responsible, professional and positioned to act as a genuine alternative in the real world. A quick overview of emerging initiatives that stand apart from large commercial oligopolistic groups and show promise for establishing journal publishing and dissemination structures in the United States highlights the role played by patrons who act as private sponsors for initiatives with a public mission. The social fabric that supports private initiatives for public benefit is not transferable, remaining an atypical phenomenon. Elsewhere, it should be up to public authorities to act to promote public services.

We currently find ourselves in the midst of a neo-liberal upsurge, with all the underlying powerlessness and fervent and incantatory credos that go with it. Since it is well known that nature abhors a vacuum, public inaction gives free reign to oligopolistic publishing initiatives and their commercialized, consumerist agenda for libraries. Public inaction is tantamount to rolling out the red carpet for the private sector to appropriate scientific communication at the expense of the public interest. The abdication by public authorities has a cost. We can only hope we will not be called upon to pay the bill in the near future.

Getting back to our main point, although we are not able to convincingly demonstrate its feasibility, observation of the situation in Canada, European countries and the European Union, to name only a few, shows that, while the political will is not always sufficient (as shown by the interruption of the Figaro project in Europe), it is nevertheless essential for mounting networked action involving the key players in responsible, non-profit scholarly publishing.

The issue surfaces in Canada in a rather dramatic way. Except in Quebec, where the political will and financial responsibility have been asserted, we are still looking for a concrete commitment to promoting and allowing journals to establish a publishing, dissemination and preservation structure on the Web. More characteristic is not only the absence of a structured public initiative but, above all, the absence of a clear message to that effect. The grey areas only advocate and encourage individual initiatives on behalf of aggregators and large international commercial groups. In the end, no one is better off.

The lack of symmetry in Quebec and Canada cannot be explained by a lack of information or by isolationist action because the main decision makers, on both sides, have always shared their viewpoints. Immersed as we have been in this milieu over the past few years, we know from experience that Canadian representatives outside Quebec have participated in decision-making leading to the adoption of a Quebec-type public policy for promoting journals internationally through digital means. To identify the underlying obstacle, we must look elsewhere and evaluate institutional, financial, political and cultural factors. It is important, however, to be future-oriented and, for the time being, what counts is the initiative of the various partners directly involved in the publishing and dissemination of scholarly documentation such as journals.

In the course of the last few years, some university institutions (with five regional leaders and institutional partners) and the individuals who are the driving force behind them, whose vital importance should never be forgotten,[2] have joined forces to establish a network that provides an infrastructure for producing, disseminating and exploiting research results, such as journals. We are very proud of what has been accomplished in this regard, not only in preparing an action plan but, most important, in developing the concept for an innovative information system. The

Synergies Project was proposed to public granting agencies in 2003. Beyond the weightier and more structuring issue of financing, the project is a concerted, dedicated commitment by the partners to implement an infrastructure for public service rooted in the academic community that will open new perspectives in our understanding of research information systems.

Our achievements at *Érudit* (www.erudit.org) over the past years have contributed to our vision of the future of scientific communication and led to the development of the concept of a multi-corpus information system positioned to exploit the principal forms of scholarly publishing and research, making cross-cutting use of documentation in its entirety, whether articles, books, proceedings, theses, preprints or research data. We have just begun to explore this territory with *Érudit* and are now, in collaboration with our Canadian partners, committed to producing the architecture that will allow us to draw on all the benefits of the idea. The *Synergies* information system will not only be an innovative and multi-functional mode of disseminating and using complex and diversified documentation in Canada, but will also include a laboratory to generate new resources and remain on the leading edge with new and original research tools in various sectors, from instantaneous translation to the sociology of knowledge and text analysis.

While we cannot provide a detailed presentation of *Synergies* in this book, we would like to refer to the concerted, collaborative effort in Canada in recent months to highlight two ideas. First, one of the main components of *Synergies* is the production of knowledge in the form of journal articles, documentation that is the main artery of scientific communication. That emphasizes the urgency even more, if need be, of acting in the field of journals. If we want them to play a role in knowledge dissemination (not only as researchers, but also as vectors), Canadian journals must equip themselves with real resources in the digital world, for the renewal of both their editorial mission and dissemination methods. This is where public action becomes necessary. Second, action has been undertaken, through the formation of consortiums, that relies first and foremost on the strategic approach to developing the network. This book deals with the vital issue of networks and highlights the different levels of relevance of their development. The language aspect is related, with the unique characteristics of language groups not dominant in scientific communication, as is the case for the French language. Although this aspect is less relevant for an English-speaking public, the topic of networks is equally strategic for all.

Our Canadian initiative aims to establish a nationwide network, calling for a series of developments that goes beyond isolated and local action and stabilizes, in institutional forms adapted to diverse realities, an information infrastructure dedicated to the exploitation of research in Canada and the inclusion of such research anchored in global networks. In short, the aim is to ensure our survival and to function in knowledge dissemination channels as a first-rate partner in global research. Furthermore, even if the language variable is added to the central issue, we should not underestimate the direct contribution of the implementation of a network of diverse language subgroups to global scientific research heritage.

An understanding of networks cannot be limited to organizational agreements, but must also show an interest in developing expertise in digital processing appropriate to scholarly documentation and initiating research activities. That also implies incorporating, often with a distributed model, diversified and multidisciplinary expertise to account for production, dissemination and integration in an infrastructure for scholarly publishing and research. We are dealing here with a team of people who cannot be reduced to a single professional profile. Our discussion on the transformation of working conditions and publishing sites applies more to the whole set of players participating in the development of the information system.

Those observations underline the impetus for writing this book now (only recently published in its original version) and should encourage us to emphasize one particular aspect or another, to introduce specific considerations and to highlight individual concerns. Moreover, the approach, analysis and prospects for action are becoming established as a core concept and proving relevant as we expand our national and international collaborations. An advanced understanding of social transformation in this sector as well as a multi-dimensional and interdisciplinary vision of the constituents of stable and institutionalized networks seem to us to be the primary antidotes for magical and wishful thinking that seeks quick fixes. Reflecting neither lack of ambition or vision on our part, on the contrary, they are the necessary conditions for harmonizing the ambition and vision to build something in real time.

Gérard Boismenu
and Guylaine Beaudry
January 2004

INTRODUCTION

An essential segment of scholarly documentation, journals entered the new digital world a number of years ago. We propose to examine the conditions for that shift in the area of the humanities and social sciences, in conceptual, economic, technical and organizational terms.

Our views have evolved through our ongoing commitment over the past five years to rethinking the digital production and dissemination of scholarly journals, as well as to experimenting with new methods and trying to incorporate that thinking into the reconstruction and acceleration of scientific communication. On a broader scale, we contend that the situation of journals effectively illustrates digitization issues in scholarly publishing.

In this book, we interpret a series of phenomena and issues not so much to defend any school of thought, but rather to arrive at a realistic assessment of the many processes that are transforming the way research results are disseminated through journals. In this way, we hope to make room for alternative approaches rather than to impose one fixed view.

Our discussion is fed by state-of-the-art research in the field in addition to our respective experiences at the Université de Montréal. With different training and backgrounds

(Gérard Boismenu is a professor and was scientific editor of a university press, and Guylaine Beaudry is a librarian and information science specialist), we have developed the first infrastructure for digital publishing for a number of journals in Quebec and Canada. That experimentation has proven both demanding and instructive. Our thinking has been inspired mainly by *Érudit*[1] and its current implementation of a platform for disseminating journals in Quebec. We have tried to communicate our reflection through conferences and writing that address specialist audiences as well as our natural partners such as librarians and journal editors. Our correspondents have been Québécois, Canadian and French. All these interactions have greatly enriched our endeavour, enabling us to assess our key proposals.[2]

Humanities and social sciences journals are at the crossroads of the varied opportunities offered by digital publishing. The growing number of journals with digital versions and the dissemination of various types of digitized scholarly documentation mean that retrieval and consultation occur under entirely new conditions that can provide not only greater effectiveness, but also continual optimization of services and functions in an unlimited space. That reality has repercussions for all who make up research networks, from authors and researchers through publishers and librarians to readers.

The new environment has a major impact on how research results are exploited and formalized in the design and writing of texts, whether articles, theses or any other type of material. Authors and researchers must acquire the indispensable skills that allow them to understand and exploit the functionalities offered by digital text processing. Authors and publishers need to participate fully in redesigning the conditions for the development and communication of research results. With the ubiquitousness and immediacy of the dissemination and consultation of articles, they can now reach pools of researchers that were previously out of reach.

Methods of using and disseminating research results have been dramatically changed by information technology, by the amplifying and multiplying effect of dissemination without borders and by the use of services that optimize digital resources. The dramatic change affects everyone – researchers, doctoral candidates, journal editors, and publishing and distribution professionals. Without a doubt, scholarly documentation has to be reconsidered by creating new modes for transmitting research results. This involves new ways of writing articles and using the available tools as well as new digital dissemination methods. The tools include a panoply of possibilities for illustrating texts, such as still images, sound, video and three-dimensional images. Similarly, we are seeing the advent of hypertext, a technology that goes well beyond assembling paragraphs that may, by chance clicks, superficially link one bit of text to another. XLINK, for example, allows descriptions of the type of links between one or more targets and one or more sources. When in wide use, XLINK will contribute to the creation of new forms of documentation.

The transition to digital publishing will ultimately be assured when, with or without a parallel conventional print version, the digital version of the text is considered primary and designed accordingly. Digitized versions will then no longer be simply the digital display of print material that was written under the internalized constraints of print media. However, there is no one way to achieve that transition, any more than there is only one rate of adoption of change. Our concern is to identify the indicators, variables and tools for implementing the transition in a way that considers the conditions for its achievement while producing significant change.

Any useful discussion on the transformation of scientific communication, specifically on the role of digital journal publication, has to be based on identifying the players involved and recognizing their respective roles. In addition, the roles of each player have to be clarified in order to define the activities that are part of the process of publishing and disseminating research results. Too often, criticizing and condemning the oligopolistic practices of large commercial groups in scholarly publishing have led to generalized judgments of all journals, including journals produced by the non-profit sector. This confusion is readily made and is practical, but should be avoided.

That is why we based our study on a characterization of the dominant journals of eight disciplines (both social sciences and natural sciences) that showed, even among the top ranks in the hierarchy, that half the dominant journals are controlled by non-profit publishers, such as learned societies and university presses. They play a primary role and their marketing practices radically distinguish them from the large groups. In addition, their journals have considerable impact and visibility. Apart from the large dominant journals in international scientific communication, we must also consider journals that participate in national communication infrastructures. Using Canada and France as reference societies, a number of similarities stand out, such as the editorial and managerial independence of journals, dispersion of publication sites, high concentration in the humanities and social sciences, strong presence of non-profit publishers and publishers with related practices, fairly modest subscription rates and financial vulnerability (Chapter 1).

A study of the configuration of players allows us to explore the breeding ground where a knowledge dissemination structure could be implanted to serve the research community. Professional scientific publishing that abstains from an unbridled profit orientation does, in fact, exist. However, it does not have sufficient weight in macro-social terms or in the generalized interpretations at the basis of some options. By discarding this reality, alternative efforts often collide with the broad, non-profit journal sector, which is not the source of the problem. Thus inadvertently impairing a participant of the publishing world is probably the result of both a misunderstanding of the real situation and the high visibility of the oligopolistic practices of commercial publishers.

The marketing practices of large groups have highlighted the issue of the costs associated with scientific communication, imposing the cost issue as a major theme in the digital world. Values and conditions for survival also associated with the digital world include immediacy, availability, proximity, individuality, interoperability and networks. On the whole, propagated notions such as "easy access" and "user-friendly" in connection with digital publishing ultimately function to erode any understanding of the professional work behind publishing and disseminating scholarly documentation, such as journals. With this backdrop, the e-publishing environment may easily come to see intermediaries as nothing more than a cost factor that is to be eliminated.

Though it may seem trivial, it is certainly not in vain to recall that using high-performance, user-friendly word processors does not constitute publishing (Chapter 2). The inability to recognize the value of work done by others may lead to treating it as accessory or believing it is dispensable or replaceable. It is important to understand the contribution of the different processes that comprise the publishing circuit, from submission of a manuscript to a journal to its delivery to readers. Assessing publishing work does not mean defending the body of publishers. We can expect the digital shock wave to affect organizational forms as well as the varied professional backgrounds at work in publishing, opening new prospects to be surveyed. It is the nature of the contribution made by publishing that matters, much more than where it is performed, the individuals responsible or even the institutional form it takes.

Digital publishing is often proposed as a source of savings and a condition for making large corpora available at a low cost or free of charge. However, anything "free of charge" does have a cost. To see that more clearly, we just have to break down the costs, the method of funding and the conditions for access to a corpus. Here, as elsewhere, matter is neither created nor destroyed. The bottom line is that free access largely comes down to who pays. Without lapsing into a detailed budget exercise, it is crucial to understand the role, activities and skills needed in the publishing and distribution of journals and other scholarly documentation.

Making documentation or a collection of articles or journals available on the Web opens up horizons that could barely be imagined a few years ago (Chapter 3). It means making them available to the world. However, the exhilarating perspective of wide-reaching availability is no guarantee that the documentation will have the influence and visibility expected and hoped for, far from it.

Challenged with information overload, users have to discern, identify and select from the flow of information coming from all sides so as not to be completely submerged. That is why there is a need for filters capable of sorting, distinguishing, selecting and channelling information according to its nature, quality, type and other criteria. An undifferentiated torrent of diverse information and content carrying everything along in its path makes it invaluable to have screening mechanisms that can filter and sift the specialized information provided by journals.

It is in this world that a distribution site can act as a mechanism to screen the torrent of information conveyed on the Web, recombining, organizing and preparing information and creating thematic units. It can also provide tools, facilitate consultation, create an environment, define a focal point and act as a structuring factor. To the extent that the site fulfils its role and mobilizes the necessary attention and resources, it is an amplifier. A dissemination platform makes participation and a visible presence on the Web possible without wholesale imposition. However, availability on the Web does not necessarily mean dissemination.

That requires a strategy and the mobilization of skills (many of which are new types of skills) to actually increase the dissemination of collections of articles, particularly to new readerships. The strategy must take into consideration the most current methods used by researchers to search holdings by subject, author and keyword rather than by reference to the name of the journal or publishing institution.

The adoption of a digital version by journals does not, in principle, raise major obstacles; nonetheless, change cannot be decreed. The journal is an institutional form that is part of a complex of institutions and a scientific community with practices that have been renewed, sanctioned, objectified and legitimized by usage. Beyond the good will and enthusiasm of individuals, this structured, and changing, set of practices presents a complex field of dimensions to consider.

For that reason, it is fairly surprising that prophetic and programmatic visions of the implementation of the digital model for transformation and, in particular, the functionalities to be generalized, can ignore technical data. To be on solid ground, action undertaken to elicit change should combine the choice of technical options (based on the outcome sought) and the social and economic conditions for its achievement. The charm of adventurism dissipates quickly when faced with the setbacks that arise from unrealistic and vague assessments of the changes (both technical and social) taking place in journal production and dissemination.

A golden path for the digital transition, which would include technical as well as action guidelines, has every chance of being illusory in many cases. Some formats and formulas may, of course, be preferred, for the best of reasons. For example, emphasizing the importance of structured language (i.e., its quality, value, potential and functions) in the digital publishing, dissemination and preservation of journals is well justified. However, that does not mean any other way is inappropriate (Chapter 4). Knowing the range of technical possibilities, particularly their features and potential, proves valuable in making the most well-informed choices. The most appropriate option is the one that allows the most effective use of digital functionalities for journals under given conditions, which often have constraints.

Constraints are not only related to skills, the technological environment and the resources available, but also arise from the social process of adopting innovation (Chapter 5). Under the circumstances, it is hardly surprising that time frames are colliding (digital time, time for social change and prophetic time) and that phenomena have their own depth and history. Journals incorporate the behaviour and expectations of many players in the publishing chain, including authors, editors, reviewers of the submitted articles, editorial assistants, technical producers, distributors, readers, users, directors of granting agencies, universities and marketers. The overall problem is specific to each of them, resulting in an expected model of behaviour. This helps us understand why many participants are faced with failed expectations despite their stated willingness.

That observation calls for realistic action that maintains the necessary tension (as is required in any social transformation) between the forces and direction of change, on the one hand, and the components and behaviours that absorb the transformation and are renewed by it, on the other hand. We are referring to a number of highly diverse aspects, such as institutional recognition of the digital version of journals and the development of a political economy, organizational forms and institutional support for group and network practices. A recipe for those factors, as for the technical factors, has every chance of rapid ageing and artificiality. It is more appropriate to determine a process for addressing the issues and proposing solutions in a dynamic process.

The issue is not limited solely to digital journal publishing, but derives its full meaning from the specific conditions in which it evolves. Journals that are characteristically part of *national infrastructures* for research dissemination, particularly in the *humanities and social sciences*, and that are part of a *non-dominant language subgroup* on a global level (such as French language journals), play an essential role in scientific communication in diverse societies. However, their lot is generally precarious. The emergence of digital publishing and distribution can be said to go hand in hand with the confirmed fragmentation among publishing sites and other forms of associations.

Bringing together the resources to implement digital publishing, once the editorial work has been done, is more likely to ensure sustainable, quality services, in step with the improved standing and influence of the journals. The accumulation of resources must come from and serve the academic community. The non-profit organizational environment is in a position to create a space where journals not yet controlled by oligopolies can establish themselves as stable and professional forms of scientific communication. The space can be built on existing structures, alongside commercial groups and preprint servers. Journals have to be able to make the transition to digital publishing based on a model that meets the needs of the environment and the type of documentation. Similarly, they have to contribute to implementing a dissemination system that ensures their viability as organs of scientific communication and institutions that bring together all the material conditions for their survival.

Many initiatives support the idea that the space exists (e.g., HighWire and MUSE in the United States). The creation of such space is the first step toward ensuring the survival of journals in national infrastructures in the humanities and social sciences. From that starting point, it is easier to identify the steps for creating scholarly publication networks. Establishing networks allows the non-competitive academic community to develop expertise in the Web focal points that form the various links in digital publication and dissemination. The attraction of establishing networks of journal sites for dissemination can readily be seen, in that a number of entry points are created to large collections disseminated at various host sites. However, that is just the starting point for the benefits that can be expected.

For the Francophone world, it would enable the creation of links for the retrieval and consultation of nearly three hundred journals and tens of thousands of online articles. A mass of material like that should make its presence felt in scientific communication, first in the Francophone community, then in the Anglophone world. Affirmation of the Francophone subgroup and its institutionalization are major steps, although not an end in themselves. On one hand, that activity contributes to reducing the relative proportion of English on the Web (in stride with a larger movement that advocates a multilingual Internet). On the other hand, it is an avenue that allows freedom of movement in the dominant Anglophone world. Combining efforts in the Francophone world produces a mass effect. The capacity to provide a collection of hundreds of journals with metadata that use the same protocol makes the international dissemination of notices for articles in databases and information systems much more feasible. The metadata would include, at least, the English translations of article titles and abstracts. Exploiting the logic and possibilities of digital publishing and networking allows us to meet the challenge of disseminating Francophone scholarly documentation on a worldwide basis. We would like to point out here that our approach is not limited to Francophone journals – documentation from any language group and from any non-profit organization can be integrated in international dissemination channels using this concept.

The challenge is particularly exciting. How do we give Francophone journals in the humanities and social sciences their full place in the major Web networks that remain dominated by the Anglophone world? On the whole, co-operative efforts seem to offer the most promising results.

1: JOURNALS: FIELDS, PLAYERS and PRACTICES

The scientific communication system and, in particular, the role played by scholarly journals are undergoing a period of dramatic change, an observation that has led many to emphasize the urgency of the situation.

In the case of scholarly journals, we may go so far as to question their viability as a central vector for disseminating research results. Before proceeding with conclusions of any kind, a careful review of the situation is required. The format and operating methods of journals are currently undergoing renewal; nevertheless, their role in validating and transmitting research results remains firmly established.[1] The problem posed by journals is also exacerbated by the inflationary spiral in journal subscription prices, creating difficulties for university and research institutions. With stagnating resources, institutions face dizzying price increases that go well beyond inflation and changes in characteristic journal parameters. That vicious circle has been thoroughly studied, and its effects are evident.[2]

We must go beyond that straightforward observation, because simplistic imagery is of no great use in understanding the journal world. In fact, the concept we have of a journal is imprecise, because the word designates significantly different

FIGURE 1 Monograph and Serials Cost in Association of Research Libraries (ARL) Libraries, 1986–1999

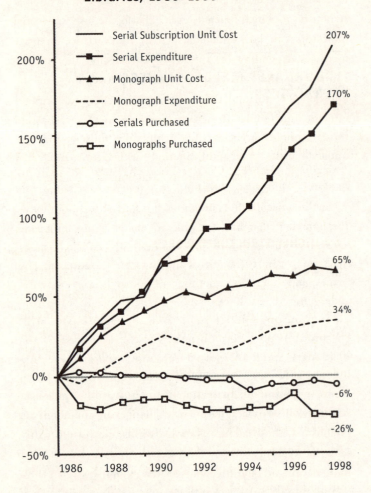

Source: ARL Statistics, Graphs and Tables for ARL Statistics, Monograph and Serials Costs in ARL Libraries, 1986–1999, <http://www.arl.org/stats/arlstat/1999t2.html>

realities, depending on the scientific community. Moreover, in considering the conditions for distributing and marketing journals, it is useful to describe the social and economic structures in which journals exist. That is the only way some indicators can be evaluated, such as the variance in price increases depending on the type of publisher, the striking contrast in the rate of price increases depending on the knowledge sector and the fact that the highest subscription rates are undergoing distinctly greater price increases.[3] Beyond those examples, it is clear that distinctions must be made to understand the scope and limitations of the scholarly journal world as well as to inspire appropriate action.

A HIGHLY STRUCTURED WORLD

The world of scholarly journals is highly structured from at least two points of view. First, scientific communication networks are institutionalized and have a hierarchy, an order of precedence and established status among journals that gives them varying degrees of validity, legitimacy, influence and impact. Communication network structuring is not homogeneous by sector. In the humanities and social sciences, where paradigms are often competing and not firmly established, the structure is more porous; conversely, in the scientific, technical and medical sectors, the structure is much more dense and definitive. Surely that explains why, in those sectors in particular, large, authoritative journals (in terms of research quality, institutional recognition and influence) dominate despite other considerations.

The structured and structuring universe is the economic and corporate organization that supports scholarly publishing. Large oligopolies play a dominant role in the production and distribution of journals, particularly in sectors where the hierarchy of scientific communication networks is most compact. According to some estimates, commercial publishers control 40% of the 6,771 scientific journals in the United States and carry considerably more weight if sales are taken into account. Learned societies produce a quarter of the journals, while university presses and public and private research organizations publish the remaining 35%.[4]

The distribution of dominant journals in the eight disciplines will be examined later. The prices set by large oligopolies allow for very lucrative profits, to the point where scholarly publications are one of the most profitable segments of all their operations. In this field, the largest concentration of business through company mergers was accompanied by substantial increases in subscription prices.[5] A pricing analysis can be found in the Wyly study, which uses 1997 as a reference year.[6] It shows that if Reed Elsevier, Wolters Kluwer and Plenum Publishing had limited their net profit margin on sales to a level comparable to the median in the serials publication branch, clients would have saved approximately US$240 million in the area of scholarly publications alone.

The two levels of structuring have cumulative effects that allow some large companies to make particularly large profits, exacerbating the financial crisis in libraries and making current methods of scientific communication vulnerable. Some learned society publishers are also able to generate guaranteed revenue for their journals, with the accompanying benefits for their organizational survival. However, non-profit scholarly publishers operating in the university milieu generally do not earn any profit and have a cost-recovery policy with a low surplus, in the best-case scenario.[7]

Sources of wealth and comfortable profit margins are foreign to most players other than large scholarly publishing oligopolies. For a clearer picture, scholarly journals must be positioned in the scientific communication system in order to highlight their specific mission. From that starting point, we can appropriately direct attacks against the prices charged and focus discussion on the cost structures in scholarly publishing. High subscription costs are no doubt due to the publication cost structure and, above all, to oligopolistic practices that pocket the brunt of these costs as profits. A comparison of the practices of different types of publishers is an excellent demonstration of that point. Such evaluations go beyond the general, inaccurate, and therefore deceptive, macroscopic level and allow strategic action to be proposed on the basis of existing structures.

JOURNALS AND SCIENTIFIC COMMUNICATION MODELS

A number of characteristics of research activities are specific to the sciences and the humanities and social sciences disciplines.[8] These distinctions explain the models developed in the past as well as new, emerging communication systems.

SOCIOLOGICAL STRUCTURING OF DISCIPLINES

The natural sciences differ from the humanities and social sciences not only in their methodology, mobility and training of young researchers, but also in their modes of social interaction and communication. These differences justify the existence of separate systems for disseminating research results.

"Invisible colleges" in the natural sciences are crucially important for the ongoing exchange of information, preprints and research reports. The high "problem-researcher" ratio creates strong competition. There is significant demand for, and supply of, frequent progress reports, which necessitates rapid and broad communication methods, through conferences (more numerous and better attended than in the social sciences), preprints and journals. Journals have shorter peer-review periods for articles, and there are some weeklies among scientific journals that are often among the most frequently cited, including *Science*[9] and *Nature*.[10] However, the preferred method of communication is the exchange of preprints, and frequent visits between colleagues allow immediate reaction and an exchange of unpublished details.

The language of the natural sciences also differs from the language of the social sciences and humanities. Hacking[11] demonstrates the inaccessibility of some scientific texts by citing an excerpt from an article on neurophysics: "If postsynaptic adrenergic neurons in neonatal rats were chemically destroyed with 6-hydroxyudopamine [...] the normal development of presynaptic ChAc activity was prevented." An uninitiated reader's comprehension of that sentence is limited to grammatical analysis.

As we move away from the core of natural sciences toward applied and social sciences, the pressure for short pre-publication time decreases.[12] Attendance at international conferences is not as frequent and serves more as testimony of social solidarity between researchers in the same discipline.

Researchers are more aware of developments in their discipline by reading colleagues' publications. It is not rare for the publication of a journal article to take a year and the publication of a book up to two years.

This overview of trends does not reveal an unchanging situation; we are seeing mimicry of the natural sciences, technology and medicine by some disciplines in the humanities and social sciences, particularly psychology and economics.

SCIENTIFIC COMMUNICATION MODELS

Well before the Internet, scientific communication was taking place based on a number of models. *A fortiori*, new information and communication technology has multiplied the models for information exchange.

Researchers in the natural sciences confirm the importance of invisible colleges and personal communication through the use of e-mail and preprint servers. Invisible colleges are being democratized by preprint servers, which radically change the rules for research dissemination. The model proposed by Ginsparg[13] was implemented with resounding success in the physics research community and reaches beyond the association of journals with a particular school or current. Nevertheless, it could be criticized for its centralizing effect.[14] Similarly, other preprint servers have been established, including servers in the humanities and social sciences.[15] The concept of a preprint may be useful in several ways. It may be defined as an article that has not been peer-reviewed – unlike the procedures followed by scholarly journals. However, it is also presented as a solution for self-archiving and dissemination by authors[16] or Skywriting,[17] referring to the free circulation of research results to anyone with access to the Web.

Le Crosnier rightly notes that "in some communities, when the urgency of publication is predominant and the scientific network not very extensive and well-structured, preprints play a central role in recognition." [18] That does not prevent researchers who distribute their papers on preprint servers from sending most of them to journals for more formal, recognized publishing. Apart from the Los Alamos example in the physics research community, "other disciplinary archives [... still only] contain a minuscule portion of the annual corpus of published papers." [19] Moreover, the creation of Los Alamos has not had a negative impact on the journal *Physical Review* even though the fields of interest of the journal and the preprints stored on the server largely correspond. [20] That means authors value scientific journals as a means of receiving validation and recognition for their work, a process that goes beyond communication between researchers. [21] Up to now, a preprint has not been considered a publication, particularly with respect to the recognition of authors by their peers and the importance of peer-reviewed publications in the process of promoting professors and researchers. In addition, quality control and archiving of papers submitted to preprint servers generally pose a problem, because the files are archived as the authors send them, without conversion of texts and data to a standardized format. [22] That is why preprint servers and journals have different and complementary missions.

Describing the history of scientific communication from letter to printed book, from book to journal article and from journal to bibliographical database, Price concluded with the observation: "At each turn, let it be noted, all the old mechanisms were preserved but new technical advances have modified the whole system and produced new reactions, new forces."[23] In this field as in many others, as major as the changes are, transitions have rarely followed a *tabula rasa* approach; instead, they took place through renewal of practices, reconfiguration of the whole and sedimentation of constituent elements. That is the complex process at work in scientific communication. The real issue is the coexistence of preprints and journals. Just because a new technology or new form of communication emerges does not necessarily mean it should replace the previous one. The communication of sound and text messages by satellite between ships' captains took over from the Morse code; however, today, we use the telephone and voice messaging without thinking that e-mail alone could suffice.

The primary objective of preprint servers is the *communication* of correspondents' research results by disseminating their papers in the state they were submitted, i.e., without filtering on entry to the preprint server. What distinguishes *communication* from *publication* is that a publication is created by a whole process of selection, processing, formatting, distribution, institutionalization of forums, recognition and archiving of the papers submitted. That is the territory of journals.

Communication facilitated by e-mail and preprint servers transforms publishing and editing models of scholarly literature from the viewpoint of both technology and distribution methods. However, as was rightly pointed out by the publisher of *D-Lib*: "Today, digital libraries and electronic publishing provide authors with many other options: ranging from departmental reports, to e-print archives and electronic journals. While it is clear that these new methods of scholarly communication are widely used, there has been little systematic study of how specific disciplines are making use of the alternatives." [24]

RANKING TOP JOURNALS

To clearly define strategies for action, first we need to identify the relative positions of the players in scholarly journal publishing. First of all, let us take a look at key international journals, and in a second step, consider the situation of journals in national infrastructures for research dissemination.

In this approach, the classification of the top ranks in the journal world is based on a review of the actual situation in representative disciplines, i.e., four disciplines in the natural sciences and four in the social sciences, which allows us to identify the publishers, gain an understanding of their commercial practices and gauge their relative weight in scientific communication channels. Data on the journals' impact factor serve as a tool for determining the journal sample and for showing the relative importance of journals in the research community.[25] It is then possible to inventory journals in France and in Canada, two countries that are not leaders in scientific networks, but participate actively in them.

Examining the characteristics of journal positioning in the disciplines involves introducing classification factors. The typology used by Becher[26] distinguishes four types of disciplinary fields, i.e., *hard pure, hard applied, soft pure* and *soft applied*. Two disciplines are associated with each type. In the first case, *hard pure*, we selected chemistry and physics; in the second, *hard applied*, electrical engineering and materials science; in the third, *soft pure*, political science and sociology; and in the fourth, *soft applied*, education sciences and social work. In each disciplinary field, twenty-five journals were selected with the highest impact factor for 2000, based on *SCI Journal Citation Reports* and *SSCI Journal Citation Reports*.

The impact factor must be used cautiously. In this context, the measure allowed us to select the journals that apparently had a dominant position in the disciplines and to select a qualified sample for each one.[27]

A simplified picture can be drawn from the domination of scholarly journal sales by commercial publishing oligopolies. The stock view is that large private groups have taken over most of the territory occupied by the best international journals, their privileged position allows them to collect particularly advantageous guaranteed revenue and public institutions, libraries in particular, have no choice but to subscribe to their commercial practices, because the journals involved are the largest and most in demand by specialists.[28] In other words, it might be concluded that everything important in the journal world has been taken over by publishing oligopolies. However, that is a generalization based on often reiterated statements that erases a significant part of the real journal publishing situation, foregoing and blocking rich prospects on the part of the research community to take control of current methods of scholarly journal communication.

Beyond the commonly held view, we have to discern the relative position and practices of players among the major international journals. From that starting point, we can then look at Canadian and French journals, which are not generally in the top ranks.

DOMINANT JOURNAL PUBLISHERS

University presses are practically absent from the science sectors, represented by chemistry, physics, electrical engineering and materials science. A single journal in the group, the *Israel Journal of Chemistry*, is published by a university press, the Wizemann Science Press of Israel. Journals are distributed as follows: 55% are published by learned societies and 44% by commercial publishers. The two groups show a slight variance in average impact factors (3.44 and 2.83, respectively).[29]

The situation is somewhat different in the social sciences, represented by sociology, political science, education and social work. The pool of dominant journals in these disciplines is distributed equally among the non-profit sector (learned societies with 30% and university presses with 20% of journals) and commercial publishers (50% of journals). The average impact factors are 1.35 for learned societies, 1.07 for university presses, and 0.88 for commercial publishers.

Commercial publishers publish half, or slightly less, of the dominant journals in the fields this study focuses on, both

TABLE 1   Relative Weight and Impact Factors by Type of Publisher

	SOCIAL SCIENCES		NATURAL SCIENCES	
	Relative Weight	Average Impact Factor	Relative Weight	Average Impact Factor
University Presses	20%	1.07	1%	-*
Learned Societies	30%	1.35	55%	3.44
Commercial Publishers	50%	0.88	44%	2.83

* The presence of a single university press is too marginal to include in subsequent calculations and cannot represent a sector alone.

---

in the sciences and social sciences. University presses, where applicable, and learned societies share the rest of scholarly publishing, which leaves plenty of space for publishers in the non-profit sector.

SUBSCRIPTION COSTS

The average price[30] of journals in the natural sciences is nearly five times higher than in the social sciences, with respective amounts of US$1,309.82 and US$276.15.[31] There is less difference between the median prices, which are US$802.50 and US$195.00 respectively, a ratio of approximately 4:1.

Librarians will not be surprised to note that the average prices of dominant journals, when viewed from the angle of

TABLE 2 Journal Subscription Price by Type of Publisher

	SOCIAL SCIENCES		NATURAL SCIENCES	
	Average	Median	Average	Median
University Presses	$145.39	$125.00	-	-
Learned Societies	$128.32	$118.00	$899.91	$625.00
Commercial Publishers	$411.25	$358.00	$1,969.14	$1,689.50

the type of publisher, are particularly disparate. Although the average cost of natural science journals published by learned societies is US$899.91, the average cost of journals published commercially climbs to US$1,969.14, thus more than doubling. Moreover, the ratio shifts to slightly less than triple when the medians are compared, respectively US$625.00 and US$1,689.50.[32]

The situation is similar for the social sciences, although the dynamics of the three groups have to be considered. University presses are much better represented in this case and have practically the same subscription prices as learned societies, with respective averages of US$145.39 (university presses) and US$128.32 (learned societies) and medians of US$125.00 and US$118.00. Again, journals published by commercial publishers have higher prices by a factor of approximately three, whether the average (US$411.25) or the median (US$358.00) subscription price is considered. The situation is thus fairly similar to the situation in the natural sciences, however with a different price scale.[33]

The relative weight of journals could be examined based on layout, complexity and frequency of issue. Some are published once a year, others twelve times or more. Some have several hundred pages annually; others are printed in small font on several thousand sheets of bible paper. The fact remains that, when comparing journals in a limited number of disciplines with similar status, the differences cancel each other out in the major fields. They regain their relevance when it comes to accurately assessing the justification for the price variance between the major fields and are likely to account for only part of the explanation. Other studies similar to ours, although more focussed on a particular field in the sciences, have produced similar results. For example, the Scripps Institution of Oceanography Library determined the average cost per page for each of the publishers in the field of oceanography for 1996 and concluded that "Society and university publishing tend to have a lower cost per page than commercial publishers but there are exceptions."[34] The situation has not changed.

FREQUENCY OF USE AND IMPACT

It is often claimed that journals published by large commercial publishers are chosen by librarians and universities because of the frequency of consultation and their importance in the field. That claim has elicited some scepticism.

Wilder, for one, contends that commercial publications represent 75% of the titles in the chemistry library at the

University of Illinois (Urbana-Champaign), but only 42% of the titles in use, as measured by the number of times a title was circulated or reshelved. Yet, with only 25% of the titles, non-profit associations represented 58% of use. A measure of citations of commercial chemistry journals Louisiana State University subscribes to shows similar results. For 78% of titles, periodicals from commercial publishers represent barely 50% of all citations attributed to periodicals that library subscribes to.[35] Wilder's conclusion is rather strongly stated, emphasizing the point that if university libraries make the health and technology collection central to their acquisition policy, they cannot escape heavy charges, particularly when low-use titles make up a large proportion of the budget for health and technology periodicals. That conclusion may be hasty, if we consider that data on the impact of chemistry journals are characterized by a particularly abrupt asymptote, which illustrates the large concentration of citations in a small number of publications. If a commercial publisher was at the top of the list, it would modify Wilder's calculations considerably, casting doubt on his conclusion.

In any case, bearing Wilder's demonstration in mind and in an effort to establish a measure for determining the relative rank of journals within the research community based on the type of publisher, we tested the generally accepted statement that the journals with the highest indicators are concentrated among large commercial publishers.

We determined an average score for journals by type of publisher. Each of the twenty-five journals was given a score corresponding to the reverse order of its rank in its field of specialization. Journals in the top ranks for impact factors in

TABLE 3   Average Score for Impact of Journals by Type of Publisher, 2000

SOCIAL SCIENCES	Average Score	# of Journals
University Presses	12.55	20
Learned Societies	15.63	30
Commercial Publishers	11.60	50
NATURAL SCIENCES	**Average Score**	**# of Journals**
University Presses	-	1
Learned Societies	13.66	55
Commercial Publishers	12.25	44

their field were thus given twenty-five points, compared to only one point for the journal ranked twenty-fifth.

As we can see, different types of publishers tend to be dispersed fairly equally around the average, both in the social sciences and the natural sciences. In other words, large commercial publishers do not show a particular advantage in terms of the relative impact of their journals in scientific communication.

WHAT IS THE PRICE OF IMPACT?

The cost of subscriptions is much higher among commercial publishers, and their impact indicator is not differentiated from the indicators for other types of publishers. As a result, acquiring periodicals from commercial publishers is distinctly more expensive than from other types of publishers.

That can be illustrated as follows. If an institution subscribed to all the journals in the sample, three-quarters of the acquisition budget in the social sciences and three-fifths in the natural sciences would be channelled to commercial publishers for journals that represent respectively only 42% and 36% of citations in the various fields. The cost of journals from non-profit publishers is evidently much more modest, given their impact.

Although only an indicator, the data reveal that, regardless of the calculation method used, journals of equal quality published by commercial publishers cost much more than journals published by learned societies and university presses, both in the natural sciences and in the humanities and social sciences.[36] The data also reveal that journals from commercial publishers are not positioned more advantageously than their non-profit competitors in the eight fields examined for the same time period.

NON-PROFIT PUBLISHERS

The distinctions between publishers' practices help give a more detailed view of the position of publishers based on increased costs and the inflationary spiral that is undermining access to scientific information. At the same time, those distinctions underline the hasty, even hazardous, nature of unfair generalizations. They also show similarities among the

TABLE 4   Percentage of Citations by Percentage of Acquisition Budget for 200 Journals in Sample, 2000

	SOCIAL SCIENCES		NATURAL SCIENCES	
	% of Budget	Citations	% of Budget	Citations
University Presses	11%	20%	-	-
Learned Societies	13%	38%	41%	64%
Commercial Publishers	76%	42%	59%	36%
Total of Acquisition Budget	$19,329		$103,680	

practices of non-profit publishers, such as learned societies and university presses.

Large commercial publishers have taken control of some major journals in the various sectors and developed a marketing strategy based on an oligopolistic position that allows them to both impose their series of journals and charge exorbitant prices. That marketing strength should not obscure the more significant, in fact, majority presence of non-profit players; it certainly also highlights the fragmentation of organizational methods in scientific publication among those players. In addition to distinguishing non-profit from oligopolistic practices, networking strategies at various levels allow the non-profit sector to be more cohesive and to more effectively counterbalance the practices of large commercial groups.

The roles played by various players are not embedded in a fixed organizational form. Publishing sites in the form of independent journals of learned societies and research institutions that orchestrate all operations, university presses and, more recently, services associated with libraries have adopted various organizational forms. In the current situation, those forms are being renewed through innovation and initiatives. Publishing and publication missions and practices are being redesigned but still occupy a central position in scientific communication. Before moving on to that, let us take a look at the features of national infrastructures for research dissemination.

NATIONAL INFRASTRUCTURES FOR RESEARCH DISSEMINATION

Discussions on scientific communication in which international and American structures are the frame of reference are particularly helpful when trying to identify the current issues and trends in this sector. However, in addition to journals in the top ranks internationally, attention should be paid to journals of societies that generally rank lower in the international hierarchy of research dissemination. A review of the specific conditions for the existence and distribution of those journals is instructive.

For illustration purposes, Canada and France are countries with significant research infrastructures. Although they occupy an enviable position overall, journals produced in Canada and France are not generally at the hub of network activities or in the top ranks in the scientific communication world, and there is nothing pejorative in saying so. A description of the situation of journals must include essential data on a number of areas: the international environment (which shapes the Canadian, Quebec and French environments), the journals' relative presence in the disciplinary fields, their organizational framework and publishing structure, the relative importance of the journals in scientific communication and their financial situation.[37] That starting point makes it easier to understand the ability to act on specific variables and the predictable impact of particular activities.

PUBLICATION SITES AND SECTORS: FROM DISPERSION TO GROUPING

A whole series of factors must be included in the discussion to satisfactorily account for the phenomenon of dispersion of journal production sites, as illustrated by the phenomenon in Quebec, Canada and France. The overall situation is difficult and generates insecurity among journals, not so much as intellectual institutions but as organizations that have to ensure their economic viability. The overall situation has specific effects in Quebec, a small society demographically speaking, where the official and spoken language is not the dominant language of international networks.

From the outset, we would like to point out the proportionately large presence of non-profit publishers in the societies studied and the very large relative presence of the social sciences. In Canada, a total of 201 major journals were inventoried;[38] nearly three-quarters of them publish in the humanities and social sciences. In Quebec, of fifty-two journals inventoried,[39] 65% are in the humanities and social sciences and 25% in the arts and literature, with minimum space remaining for the biomedical and natural sciences and engineering. In a similar survey done for France, using French national scientific research centre (CNRS) data, nearly 90% of the 150 journals identified are in the humanities and social sciences. That means, in a survey of the journal community, the primary respondents are editors of journals in the humanities and social sciences or in design and the arts.

For Canada, we know that three-quarters of journals are published and produced by small, non-professional publishers, such as university units (36%) and learned societies (38%). University presses publish 16% of journals and private publishers only 5%. Publishers of other journals could not be clearly identified. In all, at least 90% of publishers are non-profit organizations. Among private commercial publishers, Elsevier Science Ltd. and Kluwer Academic Publishers are at one end of the spectrum, and Éditions Bellarmin and Éditions Saint-Martin are at the other end; the two groups are at opposite poles in terms of their commercial structure and layout. In Quebec, the fragmentation of publishing sites and the marginal presence of large commercial publishers are striking. Considering the total pool of Quebec journals, the largest Quebec university press (Les Presses de l'Université de Montréal), with six journals, cannot be compared with the University of Toronto Press, with thirty-four journals.

On the other side of the Atlantic, in France, small independent publishers represent more than half (55%) the publishers; when counted together with university presses, non-profit publishers produce approximately two-thirds of all journals. The main journal publisher, PUF, has been considered a private publisher because of its long-time atypical status in addition to its recent restructuring. Another private publisher with a strong presence is L'Harmattan, with approximately forty journals. However, they cannot be described as oligopolistic publishers. More French than Canadian journals are part of large oligopolistic groups; however, that is not a dominant characteristic. What is undoubtedly more distinctive is that a significant portion of scholarly publishing is done by private commercial publishers that have made it their publishing mandate despite the uncertain commercial fortune that often entails.

In general, in terms of subscription costs, the pricing scale of small journals is entirely different than for major journals (and even more so for journals produced by large commercial groups). All publishers combined, a subscription to a Canadian journal in the sciences costs an average of US$175 and in the social sciences, US$40. In France, the average price of a journal in the humanities and social sciences is US$55. The subscription prices (see Table 2) for dominant journals in the humanities and social sciences range from US$118 to US$125 from non-profit publishers and US$358 from commercial publishers. The bottom line is non-profit players have clearly withdrawn from commercial pricing trends and even counterbalance the inflationary spiral.

The dispersion is shown by the distance maintained between and by publishers and journal editors. University presses certainly consider the journal sector difficult and costly and have found it more worthwhile to concentrate on the book sector, where publishers play their role more fully from manuscript selection to distribution to readers. The distance has engendered some distrust (more or less justified) on the part of journal editors who have good reason to value their editorial autonomy, yet could benefit from services and independent revenue. Moreover, interaction between publishers and journal editors can be avoided, because public and university policies have rendered it unnecessary.

In some universities, production and distribution services are available to journals either directly or through institutes, centres or faculties. Grant programs which provide funding for journals (and are thus indispensable for scientific dissemination, particularly in small societies) have been instrumental in promoting in-house production of journals. That was the case in Quebec where the Fonds FCAR (Quebec foundation funding research and research training) funded page-layout equipment and even the processing of camera-ready copy by editorial assistants, no doubt seeing it as a source of savings. Grants in the form of services are common practice in France, where editorial assistance is often provided by regular CNRS staff. In France, service grants constitute the main contribution to journals; monetary assistance is much more modest.

NUMEROUS INDEPENDENT ENTITIES

The consequence of the dispersion described above has been, in many cases, that each non-profit journal becomes a small business, handling the whole editing, publishing and distribution process on an individual basis. Each journal repeats the same operations at various sites, without economies of scale or transfer of expertise, such as preparing copy, handling page layout, negotiating with suppliers, setting up a subscription service, negotiating with an agency and organizing bookstore distribution. Fragmentation of that kind represents a cost that is difficult to assess accurately; however, it counts all along the publication and distribution chain. The ineffectiveness of fragmentation becomes more apparent when it comes to digital publishing; due to the lack of resources, quality publication and distribution of digital journals is often out of reach.

Moreover, fragmentation results in social constraints on the introduction of change, constraints that must be taken into consideration in the approach to the social conditions for the transition to digital publishing. By encouraging journal editors to invest in the more technical tasks of producing the print version (such as purchasing page-layout equipment and preparing camera-ready copy), the former policies had a structuring effect on social practices. For example, routinized production practices are likely to create constraints against establishing a combined form of service, such as a professional publication and distribution infrastructure for all journals.

Editorial teams remain fairly small and rely, to a large extent, on unpaid contributions, part-time staff and service loans provided, with or without monetary compensation, by centres, institutes, associations or faculties. They receive assured funding from public subsidies in addition to uncertain funding from the university or learned society they are associated with and revenue from journal sales. Editorial structures have become destabilized in recent years by the often cumulative effect of budget restriction policies on the part of granting agencies and universities, not to mention the drop in subscription revenue for many. All those factors have resulted in increased pressure on editorial teams, which already lacked resources.[40]

A survey of the revenue and expenditure of twenty-seven Quebec humanities and social sciences journals[41] shows that, on average, nearly 41% of the overall journal revenue comes from subscription sales, while the smallest percentage comes from single-copy sales. Financial support from the two major granting agencies represents 47% of revenue, shared almost equally between the Fonds FCAR and the Social Sciences and Humanities Research Council of Canada (SSHRC). The remaining revenue is fairly diversified, ranging from financial contributions from universities and associations to advertising revenue. The revenue calculations do not include in-kind contributions and services, which nonetheless constitute a value.[42] As for expenditure, approximately three-quarters pay for the journal's editorial and organizational structure and for preparing the first copy. Costs associated with print media (e.g., printing, binding, postage and shipping) make up the remaining quarter. Finally, the bottom line on the journals' balance sheets showed an average deficit of C$3,700. Despite the uncertain budget situation, journal editors nevertheless carried on from one year to the next (often with a large dose of imagination), which illustrates the instability of the organizations and their endemic lack of resources.

The contribution of information technology to the journal world needs to be considered in the framework of conditions that are more social in nature.

BREEDING GROUNDS? REALLY ...

On the basis of the previous comments, it may seem appropriate to view journals in the societies discussed as breeding grounds for young researchers. The implication is that the journals are minor scientific communication vehicles in terms of content, role and impact, a view that is unjustified, at least as far as the role the journals play and the quality of their content.

The least bit of attention shows that journal editors actively aim to practise the highest standards of quality in their editorial work, to position themselves at the leading edge in their discipline or field internationally and to participate in key international forums. Indicators of the impact of articles published are certainly lower than for large international journals in the same sectors. Moreover, the language issue is not negligible for French-language publications, casting doubt on the validity of impact indicators. Without elaborating on the issue, we can point out that researchers in the humanities and social sciences continue to publish in their "national" journals even after they have begun publishing internationally. In other words, the national journals are not being deserted by established or rapidly emerging researchers. The main issue for authors is to weigh the choice of vehicle for disseminating research results and how to reach the target audience most effectively. Young researchers proceed the same way, readily giving preference to "international" organs of distribution. At the same time, authors from other countries are frequently read in national journals.

The qualifier "national" to describe a journal is inadequate. The interest many journal editors show in new information technology is fairly revealing of the expectations they have for their journal. They see technology as an opportunity to renew their scholarly journal and to increase its influence, considerations that are not necessarily typical of regional or national mandates. That interest usually goes along with a determination to augment the journal's status and influence and a willingness to use whatever means are necessary to achieve that goal. National journals are thus aware of the essential role their journals play in scientific communication, an awareness that explains their interest in utilizing the best methods for increasing the dissemination of the journals.

DISSEMINATION: AN ESSENTIAL ROLE AND PROBLEM

Many national journals already have a long history (long in relation to the history of the scientific community), looking back on twenty-five, thirty-five or even fifty years in Quebec and Canada, and often more in France. They are major institutions in the scientific community structure, have a reputation for thoroughness and provide a setting for the validation, legitimization and recognition of research results. With few exceptions, they do not occupy a dominant position in the global scientific communication networks, from two standpoints. First, they do not occupy a top-ranking position among the most influential journals in their respective sectors and, second, they are not part of large international commercial publishing oligopolies. Despite those limitations, they manage to have a far from negligible presence beyond their borders.

Three surveys illustrate that point for Quebec and Canadian journals. Similar surveys could be conducted for France and would show the role and influence of journals as a vehicle for the dissemination of scientific knowledge.

First, a recent study has shown that, for the period 1991 to 1996, Quebec authors who published in any of the fifty-two journals inventoried in Quebec were a minority (46%). As a result, authors from other countries (37%) and non-Quebec Canadian authors (13%) occupied a majority position in Quebec journals.[43] The authors' origins varied a great deal; three-quarters of contributors came from ten countries. France was clearly the most common country of origin, followed by the United States and Belgium (roughly equal), then Great Britain and other European countries. Non-Quebec Canadian contributors were mainly from Ontario, in particular from the University of Ottawa, University of Toronto, York University, Queen's University and McMaster University. Moreover, one quarter of all articles were written in collaboration with authors outside Quebec and even more with authors from other countries. This is a growing phenomenon. Of course, established Canadian (Quebec and non-Quebec) researchers publish in other countries, but they continue to publish in national journals. Although the journals attract researchers from other countries, they have not been deserted by "national" researchers.

Quebec journals, which are published mainly in French or devote a large amount of space to French, are major players in the dissemination of knowledge, mainly in that language subgroup of scientific networks, which in turn are known to be more polymorphic in the humanities and social sciences. We would hope for more sustained dissemination of such publications in the Francophone world. Regardless of the intrinsic value of the journals, their dissemination faces the constraints and obstacles of traditional commercial networks. That should not lead to underestimating the relative importance of subscriptions from other countries (mainly from institutions) and their overall relevance for specialized communities. In the Quebec journal sample, more than 30% of subscriptions come from other countries, comprising slightly less than half the subscription revenue.[44] Given the challenges and pitfalls currently encountered by journal editors and publishers, the relatively high level of institutional subscriptions from other countries is testimony to the quality of editorial work of Quebec journals and their participation in key international forums in their respective disciplines or sectors.

Finally, in a not yet published study,[45] we have attempted to assess the presence of Canadian journals in research networks in the United States. A sample of fifty-three journals was selected, and twenty-four large research libraries at American universities were studied. The survey shows that, on average, the surveyed libraries subscribed to 30% of the fifty-three Canadian journals. Moreover, more than half the libraries studied subscribe to slightly more humanities and social sciences journals than to natural sciences journals (a ratio of 12 to 10). Moreover, half of all the Canadian journals surveyed (more than 200 journals) are listed in *Current Content* and listed in an average of fourteen abstracts, not counting *Current Content*. Quebec journals were not distinguished from other Canadian journals, but we can assume that they would be within the normal range, particularly if dissemination indicators are considered that appropriately recognize the language of communication used, French.

The distribution of print versions of journals, as instruments of scientific communication, faces barriers created by conventional commercial network operations. A change to digital dissemination could circumvent those constraints and barriers (without necessarily having to eliminate the print version), in addition to increasing the journals' presence in scientific communities abroad. Naturally, a journal's international impact must ultimately rely on the quality and universality of its content and cannot be achieved exclusively by implementing technical steps. At least in the case of Quebec journals however, as mentioned above, the quality has already been acknowledged by the relatively high number of institutional subscriptions from other countries, which also attests to their relevance for specialized communities. For journals operating in "national" frameworks, dissemination is one of the major challenges for the next few years; they must work hard to achieve increased impact in international scientific communication networks, particularly in the French language subgroup, and electronic publishing is an essential and unavoidable channel.

2: PUBLISHING in the DIGITAL ERA

Misunderstanding the publisher's role often contributes to underestimating the importance of that role. Those who would reduce the publisher's role to an intermediary between author and printer, for example, tend to see new publishing techniques as questioning the publisher's role and also tend to associate the publisher with negative experiences, such as publication delays and increased costs.

Some also argue that the main, traditional functions of a publisher are essentially obsolete and, similarly, that libraries represent unnecessary costs that can be eliminated.[1] The "unnecessary" library costs, incidentally, are distinctly higher than publishing costs. Chartron and Salaün, who have a good knowledge of the role and services delivered by librarians, discredit claims of that kind as over-simplifying the social and cognitive functions of libraries. Similar reservations can be expressed about the publishing process and its role. However, the fact remains that, no matter where or in what organizational framework the process is set in action, the publishing process is fundamentally affected by information technology, in content preparation, production and distribution.

THE PUBLISHING PROCESS

Publishing covers a broad spectrum and requires the concerted action of various participants. The main stages of the process are preparing copy for publication (corpus, data, iconography), formatting, determining the media and physical features, dissemination and ensuring that the target readership is reached. Journal publishing operates under the same division of labour, yet has the added component of delegation of the work of selecting and preparing copy to the editorial teams that produce them. Although publishing has become a professional activity, it is not the exclusive domain of players in the trade and industry. Current information and publishing technology, in addition to the traditional presence of independent agents, impede a professional monopoly (although paradoxically publishing houses do tend to form associations and mergers).[2] Particularly in the humanities and social sciences, journals are often published outside professional publishing houses. In that context, as a set of tasks to produce a final product, the publishing process is much more important than the organizational framework that shapes it. That is the process that has to be defined for the digital era.

VALUE ADDED BY PUBLISHING

With the opportunities made available by digital publishing, numerous initiatives have been developed to exploit the enormous potential of digital communication technology without necessarily having to resort to professional publishing procedures. In this way, it has become commonplace for authors to provide free access to their work. However, the *vanity press* publishing model does not always satisfy the demands for standardization and recognition of the publishing process.

Publishing provides relevant and significant value added in the production of print material and in the digital environment. Scholarly publishing, particularly digital scholarly publishing, refers to the process of institutionalizing scientific exchange forums. This process involves a series of operations on a number of levels.

The conventional role of a scholarly publisher has four main functions:

1. *Manuscript evaluation and selection*. In the case of journals, that task is delegated to the editorial committee or editor. Nevertheless, the publisher remains the guarantor of quality and ensures that the editorial committee effectively fulfils its responsibilities.

2. *Text processing and formatting*, activities that involve both copy construction — editing work in the sense of "preparing text form and content"[3] — and proofing the written product. Added to that is the physical presentation of the publication: thorough presentation, consistent application of a typographical template, the quality and relevance of the choice of material and flawless formatting.

3. *Distribution, promotion and sales*. The publisher handles marketing, promotion and distribution to ensure authors reach the greatest possible number of readers.

4. The fourth function concerns *preservation*. Conventional publishers have to maintain a complete inventory of their publications to meet the demand for previous, even out-of-print, issues. Print-on-demand offers a solution for that situation, while the actual archiving of the inventory is assumed by national libraries and the research library network and through the distribution of copies to many different geographic locations. Preservation for conventional publishers thus essentially comprises maintaining their own organizational or company archives.

The main functions of publishing in its conventional form, individually or combined, comprise the role of the publisher. They cover the process from the production of written material as an intellectual creation to its dissemination to readers. With the adoption of information technology and the development of digital publishing, the tasks remain the same, although they are clearly enriched. We should also add that they are being redesigned and that new players are needed to fullfil them. Thus, digital publishing entails a new alliance of skills and redesigned expertise, sites and organizational methods – all changes that shift away from the publisher's traditional image. The generic term *publisher* thus has to be understood as a "changing character with many faces" to embody publishing in the digital era.

DIGITAL ENHANCEMENT OF PUBLISHING

Once the manuscript evaluation and selection stage has been completed, publishing requires technological choices, particularly in formatting, processing, production and distribution. The publisher must also participate in indexing, adding hypertext links, developing a communication space (comments, discussion lists) and applying multimedia (image, sound, video). Although the development of a multimedia infrastructure can or should be the publisher's responsibility, authors greatly benefit from understanding its use in order to fully exploit data and to present their research more effectively. Publishers may rightly be accused of underutilizing the potential of digital publishing; however, they cannot, even in the best-case scenario, substitute for the author in designing and writing material.

In addition to adapting their traditional methods for Internet dissemination, publishers also have to continually develop Web services to facilitate dissemination and to optimize the use of online material. Installing a high-performance search engine is indispensable, and considering other services, such as the use of distribution lists and selective distribution of information, is strongly advisable. Moreover, because the publisher is no longer selling a tangible product but providing a service or access to documentation, the stability and quality of access to that documentation must be ensured.

Finally, publishers cannot escape their responsibility for the preservation of online material. The longevity of the information published and disseminated has to be ensured, and arrangements have to be made to guarantee long-term access to the service provided. Those requirements greatly exceed traditional practices for the management of "organizational or company archives" and confront publishers with the challenges of preservation and even archiving.

It is useful to review some of those considerations, starting with a simple observation. When we consult digital documents, we are initially struck by the lack of physical distinctions between types of material, distinctions that are immediately evident in print documents. The lack of physical distinctions among digital documents does not allow us to distinguish an article from a periodical or an encyclopedia *a priori*. The screen remains, by and large, the only interface between the reader and the information, whether it is a newspaper, journal or database. Often, that also creates the illusion that digital documents are easy to process and produce. However, discussing digital publishing without mentioning features such as coding format, navigation tools, upstream processing chain and search potential is like discussing print publishing without referring to paper quality, typographical template, type of binding and page numbers.

Digital publishing features become evident as soon as we go beyond simple on-screen consultation and attempt, for example, to use the search potential or re-use data. Just as for conventional print publishing, the publisher must make technological and artistic choices that fit the characteristics of the material to be published and distributed. The characteristics of journals, for example, comprise text, often in several languages in the same document, tables and complex mathematical formulas, black and white and colour iconography (graphics or photographs) and the use of hypertext and hypermedia.

The importance of this type of information for the research community and society imposes technological choices that allow preservation for future years, even for future generations. The selected coding format must also allow searches that eliminate noise and silence as much as possible. To meet digital scholarly publishing needs, it is important to distinguish between production and preservation formats (e.g., XML) and dissemination formats (e.g., XHTML and PDF), while ensuring long-term digital document preservation. Therefore, although implementing quality digital publishing and distribution does not have to be inaccessible and frighteningly complex, digital publishing does require professional expertise and abilities, just as print publishing does.

IDENTIFYING NEW PUBLISHING TASKS

Although digital publication and distribution technology continues to maintain, and has numerous analogies with, the conventional publishing trade and skills, it requires a complete review of publishing activities.

1. Establish a digital processing chain for the production of *by-products*. The most valuable impact of digital publishing is indisputably the fact that a publisher's products can be presented in a number of complementary forms.

2. *Digital dissemination methods.* Publishers no longer provide a tangible product; they now provide a service, in that readers now buy access to services. In this way, readers avoid having to purchase and maintain the equipment necessary to store digital publications. Preservation and indexing – functions performed up to now by librarians – ultimately become the responsibility of the publisher. The shift in the boundary between roles encourages alliances for sharing expertise.

3. *Dissemination and marketing methods* have to be reviewed and reconsidered due to the different nature of digital production, which constitutes the third task. Major changes in current activities and practices have to be planned to ensure active distribution and to reach a readership that was difficult to reach before.

4. The fourth task is undoubtedly very exciting in that it involves participating in the *creation of new knowledge transfer models* that benefit from interaction between readers and authors and the use of multimedia.

Two complementary comments arise from these challenges. First, dissemination is an activity of primary importance, although there is often an automatic tendency to take it for granted with digital communication. In print publication, it is understood that it is imperative to pay close attention to dissemination and distribution once the copies have been printed. Otherwise, we realize that large inventories often signal upcoming closure of operations. Digital dissemination does not happen automatically; being available to the world on the Web does not mean being disseminated. As a result, new expertise is needed to implement and administer digital dissemination. Most important, however, is defining the framework and conditions for ensuring the dissemination and influence of digital journals (the subject of the next chapter).

Much of digital publishing up to now has consisted of a simple transposition of documents designed for presentation in print media. From our experience and from what is available on the Web, we cannot help but conclude that the numerous possibilities offered by digital media have been underutilized, features such as full-text indexing, inclusion of active data, creation of hypertext links, open forums for comments on articles and the use of multimedia resources (image, sound, video, 3D simulation).[4] That is a fairly striking paradox because, in addition to the Internet dissemination method, they are particularly attractive possibilities generally pointed out to highlight the appeal and enormous power of digital publishing compared to print publishing. Therefore, the very possibilities that constitute real value added[5] too often remain virtual. Publishers have the important function of creating new models for organizing and transmitting information to exploit the potential of digital dissemination as much as possible, beginning when the author creates a manuscript.

The key challenge in publishing is not so much the technical production of a digital publication process, but the introduction of tools to optimize the quality of production, editing, preservation, indexing and, finally, systematic distribution and related services.

DIGITAL PUBLISHING AND THE COLLAPSE OF THE PUBLISHING COST STRUCTURE

Publishing in the digital era, far from being relegated to the rank of an ancillary activity, poses particularly exciting challenges. Unless we take a serious interest in digital publishing, it is hard to reconcile those requirements with the notion that digital publishing reduces the cost of publishing journals. However, the introduction of digital methods for publishing and disseminating scholarly journals is often seen as both an opportunity and a way to radically reduce production costs. It is often argued that digital publication alone, which involves the elimination of the costs associated with printing and distribution of print copies, brings production costs down to nearly zero and allows free access. However, that claim is very misleading and deceptive.

That argument shows a lack of awareness of the distribution of publishing costs for a journal, the nature of publishing work (whether print or digital) and the level of quality and service in digital publishing itself. It is important, however, to have both a practical understanding and an awareness of the subtler issues involved in digital publishing costs. It seems to be assumed that eliminating paper from the publishing process should greatly reduce costs. In fact, if we ask what the percentage reduction in printing costs is for an exclusively digital journal, the answer is almost invariably 100%. However, at least two questions arise immediately. What is the proportion of printing costs in the overall budget (including editorial work, publishing and distribution) of a print journal? Are digital publication and distribution exempt from their own costs? A discussion of these factors follows.

Our first estimate of the savings generated by digital publishing is an assessment of the costs directly related to the print aspect of publishing, e.g., paper, printing and distribution of copies. However, as fixed costs or the cost of preparing the first copy dominate the cost structure for journal production, the reduction in print costs generated by digital publishing remains relatively negligible.[6] Moreover, the savings are partially offset by the costs of digital data storage, software and higher paid staff. Certainly, digital publishing and dissemination costs are lower than print costs; however, the production costs represent a relatively low percentage of total costs.[7]

The comparison is, in any case, difficult to establish because, when we refer to digital publishing, we are not sure what we are talking about. Without specifications, even general ones, it is highly speculative to evaluate the costs of digital publishing. For example, publication and dissemination models as well as coding format (production, distribution and archiving formats) vary from one publication to the next. Different costs thus have to be assessed depending on the choices made, which are normally based on the nature of the documents processed and the value added in publishing. That serves to illustrate the approximate nature of estimates, if those considerations are ignored.

Looking at the issue from another point of view, the estimated relative costs specifically related to print publication, to the extent that they are cut back, create room for potential savings. In this area, we know that for print publishing of journals, the costs for preparing the first copy represent the bulk of the journal's total operating costs, 70% or 88% depending on whether we accept the estimate of the American Chemical Society[8] or the University of California Press.[9] That confirms the Quebec journal survey, which estimated the budget portion for preparing the first copy at 75%, on average, an estimate that appears consistent. Consequently, without adding the costs involved in digital publishing, we would imagine that the largest reduction in costs, considering current standards in scholarly publishing, would be at best 25% or 30% of total actual costs, because the cost of the first copy is roughly the same for the digital version. In addition, new costs specifically related to digital publishing also decrease overall savings.

Still, some[10] assume that costs can be reduced by much more than that margin and, consequently, that subscription prices to digital journals should be much lower than what is currently charged. That assumption would only be realistic on the condition that the entire publication process is revised downward in favour of the introduction of digital methods of text production. Revision of that kind is based on a minimalist concept of scientific publishing that fails to take into account characteristics specific to publishing work. In addition, a redesign of the peer-review process is often advocated and current procedures for the selection and quality of scholarly publications questioned. The minimalist view is pursued logically as far as dissemination design.

Clearly, choices have to be made to ensure that costs are distinctly lower with the lowest cost structure. One choice is to radically reduce the quality of the publishing process and to stop investing in the functionalities offered by digital publishing. Another choice is to radically revise peer-review procedures and publishing, distribution and marketing standards. Although the oligopolistic prices charged by the commercial scholarly publishing sector are understandably abhorrent, we cannot help but ask how far the minimum can go.

HOW FAR CAN THE MINIMUM GO?

Based on the positions defended by two supporters of radical cost reduction obtained through digital methods in journal publishing, we can identify the resulting conditions for journal production, dissemination and preservation. In this section, we present the thinking of two proponents of the "minimalist" view, Hal R. Varian and Andrew Odlyzko.

First, Varian[11] starts from the principle that the journal production process has to be redefined to reduce staff costs. Copy preparation, including manuscript evaluation and reworking, results in costs for communication, but mainly for co-ordination, which primarily represents salary expenditures that can be reduced. Authors adopting the Adobe PDF format right at the start of manuscript preparation, followed by digital dissemination of the article (which would be the author's responsibility), could result in cost reductions from reduced staff, mail, photocopying and other expenditures. The aspect of editorial work involving text processing and formatting is largely sacrificed, because copy editing is too costly and because they are not convinced that professional-quality presentation of the articles is really necessary. Perhaps, it is suggested, only the major texts published in a year should be published professionally and benefit from new processing and special delivery.

According to Varian, shifting the responsibility to authors for page layout and typographical presentation of copy reduces the quality of journals, but he seems to think that the value added by professional editing is not really worth the trouble. Added to that is the major task of preserving digital files of the journal. Although of interest, PDF is a proprietary format and, despite recent developments that allow compatibility with XML, is first and foremost intended for printing, not preservation, because it offers no guarantee of longevity. That aspect proves to be a major concern for journals. Varian acknowledges that and points out that for preserving as well as for searching and manipulating the components of articles (such as abstracts and references), it would be better to use a structured text markup method, such as XML. That option was rejected, however, because it was considered too expensive and cannot be shifted to the authors themselves. The issue of article longevity and archiving is left unresolved, as well as the issue of making optimal use of the potential XML offers for exploiting and disseminating texts. Overall, it is assumed that all those measures, which eliminate a large part of current procedures and standards, would reduce costs by half.

Odlyzko[12] asks how far costs for a digital publication can be reduced. He reviewed the achievements of Paul Ginsparg at Los Alamos with keen interest, because he saw in them an example of extensive minimization of costs. Yet, are we to take journals back to the preprint state? Odlyzko does not go quite that far, and even tries to raise standards, although parsimoniously, because the issue is whether it is possible to reduce costs by opting for a digital version only. With the number of free online journals growing rapidly every year, Odlyzko also questions to what extent a system of free journals can ultimately be viable and occupy a dominant place in scholarly publications. Without a sponsoring agency or a source of *ad hoc* funding, it may be doubtful. It would be necessary to rely on the contribution of authors for text composition, peer-review procedures by electronic communication and the voluntary contribution of university or other staff to ensure the survival of such journals. Additional editing and administration work is often done by university staff if the journal is very small; however, it is generally difficult to maintain that form of contribution and organization once journals expand. Moreover, we might ask whether researchers have better things to do (such as research, writing, teaching and conferences) than to try, however imperfectly, to transform themselves into publishers.

Therefore, we repeat the question: How far can the minimum go? Reference to a concept of "acceptable quality" serves here as a non-specific, shifting threshold. Ultimately, the disparity in production costs is linked to different visions of what is essential and what is secondary in publishing. The two commentators agree that a journal should be offered as a product with whatever can be achieved within the available resources. With that as a priority, it is secondary to provide a journal with the features necessary to fulfil its role under ideal conditions (such as ensuring the quality of the content, editing, preservation and dissemination). Nevertheless, we cannot ignore the fact that this discussion is being conducted without the minimum information needed on digital format, information longevity and dissemination conditions. Those aspects are essential, however, if we want to establish minimum standards and assess the resources needed; otherwise the discussion can only be speculative.

The two scenarios of Varian and Odlyzko are essentially used as examples here. Practices and procedures are not fixed or carved in stone, far from it. Major transformations can be anticipated and hoped for in favour of the introduction of digital publishing. Yet the scenarios serve to illustrate a minimalist view of scientific publishing, arguing that knowledge and research results (in and of themselves fairly expensive to produce) should be disseminated under the most economical conditions possible, at the risk of yielding to an erosion of standards for publishing, access and data preservation. Little importance is attached to editing, including aspects specific to digital publishing, such as production formats, preservation, dissemination and the use of multimedia functionalities. Digital publishing is reduced solely to its capacity for broad availability at low cost.

In the same vein, the problem of preservation has been raised and remains unanswered. It is imperative to ensure the longevity of the information contained in journals. Print publications that are distributed and preserved in a number of libraries and generally subject to compulsory legal deposit benefit from well-established procedures that not only ensure their longevity but also their archiving. One study shows that 15% of the articles read by university researchers have been in existence for more than five years, and 5% of readings are articles more than fifteen years old.[13] The research we have done on dominant journals worldwide (see Chapter 1) shows that the half-life of journals many times exceeds ten years, revealing the value of journal content over time. That information should convince us that the long-term reliability of publications and their canonical archives merit an appropriate response in the new environment of digital publications.[14]

Some attribute responsibility for preserving canonical archives to the publisher.[15] Moreover, the publisher is expected to fulfil that responsibility by providing practical, institutional but also technical solutions in the structuring of the processing chain. However, as mentioned, the issue is rarely raised in discussions that aim to reduce publication costs at all costs. Yet to disregard the problem is hardly compatible with the proponents' claimed interest in quality scientific publication. Conversely, the responsibility the publisher is expected to take requires explicit consideration and an appropriate response as soon as technical choices are made.

NEW MISSION, NEW RESOURCES

It is of primary importance to return to the issue of the production costs of digital material. We have seen that the hypothetical savings margin from digital publishing is much smaller than frequently assumed and that where we do see large margins, it is at the expense of eliminating quality editing and distribution work (not counting the features of a secure and reliable technical environment). In this section, we try to identify the costs associated with quality digital publishing. The picture drawn is supported by examples and estimates, in addition to reflections from scholarly publishers.

Although digital journals may be seen as a panacea for the subscription price crisis, commercial and non-profit publishers have had to, and will have to, invest significant funds in experimentation with digital publishing, particularly because costs have proven distinctly higher than anticipated.[16] The American Chemical Society estimates that the cost of the first copy of a high-quality digital journal represents 82% to 86% of the total production costs of a journal, and that the costs of a CD-ROM journal are 25% to 33% higher than its print version due largely to additional expenses for operating and search software. That estimate is supported by the experience of university presses. At MIT Press, Janet Fisher states that the overall publication costs of the digital-only *Chicago Journal of Theoretical Computer Science* are comparable to the costs of print scientific journals. Indirect costs and marketing expenses represent two-thirds of the journal's total production expenses, although the same factors account for only one-third of total production expenses for print journals. The additional costs then offset the savings obtained through digital publishing.

In the same vein, King and Tenopir,[17] at the University of Tennessee School of Information Sciences, developed an economic model to show that, for a large international journal in the natural sciences, switching to an exclusively digital journal would reduce costs by approximately 2%; the savings would then be partially offset by the cost of digital preservation, software and the highly qualified staff that would need to be hired. A switch to publishing parallel versions of the digital and print journal would cost 3% to 8% more than for the print version alone. Tim Ingoldsby, Electronic Publications Director at the American Institute of Physics (cited by Tom Abate[18]) estimated the cost increase for publishing both print and digital versions at between 10% and 15%. Furthermore, King and Tenopir maintain that the wide range of services and functionalities that comprise the value added of an entirely digital journal are likely to increase the publication and dissemination costs of the journal. Beyond the specific percentage changes, it is important to consider those expense components.

On a more local level, Sandra Whisler,[19] Assistant Director of Electronic Publications at the University of California Press, questioned the idea that digital publishing costs are substantially lower; she is referring to journals in the earth sciences and astrophysics as well as in the humanities. Below, we outline the main resources which she has identified as essential for establishing and developing quality digital publication and dissemination in a non-profit organization.

- The University of California Press postulated that development costs for the digital publishing structure could be largely assumed in the initial phase and that additional expenses could easily be absorbed by the savings generated from the elimination of the print version. That hypothesis appears highly imperfect, because there is no learning curve, but rather a succession of curves, due to a series of factors.

- Digital publishing is a field dominated by innovation, and technology and Internet use and practices evolve very rapidly, regularly raising issues about digital publication and dissemination expectations and forecasts as well as interaction with users and readers. To avoid becoming fossils, it is imperative to invest in applied research in publication and experimentation in Web dissemination, in addition to maintaining a technology watch. That requires the resources for highly qualified staff.

- The *quid pro quo* for positioning at the leading edge of technology is to consider the indirect expenses as intrinsic, rather than as incidental or optional. These expenses are indispensable for stabilizing the production process and for developing the very services that attract interest for the publication.

- The mobility of highly qualified staff should not be underestimated, nor should training time and learning costs. There is a relative shortage of qualified professionals in information sciences; job opportunities are plentiful and solicitation by companies is often pressing.

- The rapid evolution of software and the need to upgrade requires information management that should not be underestimated. Lack of information and shortsighted technological choices may prove costly.

- New qualifications must join forces to successfully produce and distribute digital scholarly publications. New responsibilities are emerging (without necessarily eliminating previous ones), such as relations with suppliers, digitizing and preserving published images, preserving digital documents, new problems and management of rights, new marketing skills (negotiating licences with libraries and institutions for digital documentation), dissemination on the Web and more conventional direct sales, as well as managing transactions and filtering online access.

However, the volume of resources necessary for digital journal publication and dissemination (based on publishing standards appropriate for the nature of the corpus processed) fails to endorse or justify the prices set by commercial oligopolies that publish scholarly research. Their pricing practices are a case of commercial exploitation maintained through oligopolistic behaviour. This problem must be dealt with in an entirely different framework, because the prices charged by non-profit publishers, even for dominant journals, and even more so by publishers in national scientific communication infrastructures, are not the cause of the inflationary spiral and financial crisis experienced by research libraries.

We should avoid the hasty and faulty generalization that scientific communication by journals in the humanities and social sciences (and also in the natural sciences), in the non-profit sector in particular, is only possible if exorbitant prices are charged. Similarly, we should be wary of the magical thinking that information technology could, as if by enchantment, dispense with costly practices and procedures. Nor should we be prone to believe that these new technologies can be imposed to unreservedly transform the procedures and practices of the scientific publication framework from top to bottom. In spite of the simplistic view commonly held, the cost structure of digital journal publication and dissemination is nowhere near zero.

3: DISSEMINATION
and the FREE ACCESS ISSUE

Dissemination is an essential, even strategic, component of digital publishing. The simple presence of journals on the Internet does not guarantee adequate visibility and influence.[1] For that reason, it is advisable to develop dissemination and implementation strategies appropriate to the virtual environment and to back them with adequate human and financial resources. Three major components of the dissemination process can be identified: the accessibility and ergonomics of the site, the dissemination formats and services offered and the strategic action to position the site on the Web.

A WEB INTERFACE FOR DISSEMINATION

Optimization of the dissemination of a journal or collection of journals depends on a number of factors including, at the outset, the attention paid to Web site design. Accessibility is one of the features that makes a site an effective communication tool for research.

Accessibility depends on the information communication environment. The architecture of the site and the organization of the information should aim to fulfil three missions:

1. To propose varied methods of access to information, e.g., navigation, full-text and field searches (title, author, description, key word, etc.) and document classification,

2. To exploit the power of hypertext and

3. To profit from the functions offered by the site.

The ease with which the information can be consulted is value added to the site. The care taken in the ergonomics of the interface (e.g., graphics, visual identity and simplicity, upgrading informational content) and the power of the navigation aid tools also play a full role.

Site design should build on the accessibility of informational content and the consulting conditions. Both aspects contribute to ensuring wide dissemination, to the extent that they enhance the richness of the content and formatting. Yet content alone is not enough; we also have to ensure that the site has maximum visibility for potential users, particularly by ensuring it is inventoried by search engines.

DISSEMINATION SYSTEM

A site's dissemination system provides a number of services that may be implemented over time based on priorities and available resources. In order to exploit this potential as much as possible, it is useful to identify the various services and to establish a program of activities.[2]

Some basic services are required at the outset. The reader must be able to navigate easily and always understand what can be found on the path chosen. The challenge of the navigation system is to correctly model the various descriptive records and to produce them automatically from the fundamental information units comprised by the journal articles. Thematic navigation should also be offered, which can be made possible by the use of metadata[3] associated with the articles. That allows navigation based on concepts, such as subjects, authors and thematic relationships, rather than on journal or issue. A consultation could then start with a descriptive record for an article that is the entry point to the article. The descriptive record must contain a link to all available versions of the article; once the link has been activated, the reader can consult the article at will.

However, navigation alone is not enough when the collection of documents is fairly large, which is likely for a collection of journals. The site needs a sophisticated search engine that can satisfy two essential concerns. First, the search engine must allow varied and complex search queries. The search tool must be able to search not only the text of articles, but also the metadata and the document structure; for example, it must be capable of searching for words only in subjects, authors or section headings. Second, the search tool must be able to display the results clearly and practically for users.

The majority of search engines produce results for queries submitted by consulting previously generated indexes. This type of tool gives fairly good results, particularly as it allows, in addition to full-text searches, the use of elements of the document structure (e.g., abstract, title and bibliographical references) as search criteria. Another type of search engine that is generating a great deal of interest and merits particular attention for collections of journal articles is a search tool with algorithms that perform linguistic analyses to locate and isolate concepts and semantic units of text.[4] Those possibilities coupled with structured searches allow maximum use of the data collections.

Managing filtered access becomes necessary when restrictions are imposed on consulting the material or users are offered personalized services. A user-management system entails the development of an access to information policy, the identification of information units and services that could be subject to restrictions and a definition of what constitutes a user. As we will see later, it also implies the mobilization of resources not only to implement, but also to maintain filtered access.

Other services may be added to those central services to enhance consultation and use of the documentation distributed.

Selective dissemination of information is a relevant service that meets researchers' needs. It involves e-mailing information on new publications to interested individuals based on their previously defined profile of information preferences and needs. This personalized service involves gathering user preferences for themes, authors, journals or any other criteria, in addition to offering users privileged access to "first-runs" that match their preferences. In that way, readers could, for example, receive articles by e-mail in their areas of interest and also, through cross-references, articles that cite the text of a given author.

Functionalities to assist readers may also constitute value added that is generally highly appreciated. For example, *annotation* allows readers, who are often accustomed to doing so, to take notes when consulting scientific articles. *Bookmarks* allow users to mark specific places in an article and to quickly go back to them at will. Finally, *links* created by the user are a valuable tool for knowledge acquisition in a document corpus. Added to those three functionalities is *manipulation* of images and other multimedia content, such as modifying images to highlight details as well as starting and stopping video sequences.

WEB DISSEMINATION STRATEGIES

Referencing and strategic positioning of a site in various Internet search engines, such as indexes and directories, proves crucially important to attract users.[5] Dissemination of journal content also requires visibility in the key databases available to individuals searching for information in the social sciences field. It is useful to identify databases and resellers, to assess the representation of journals in them, to analyse the potential of databases, and to develop strategies for ensuring an increased presence in such directories. Finally, a Web marketing campaign could prove worthwhile to attract new audiences, to retain the loyalty of current users and to adapt the site's services to new trends.

REFERENCING AND STRATEGIC POSITIONING

Massive numbers of Internet users (85%) use search engines to find information on the Web. Of these, over 85% do not look at more than two or three pages of the results of their search query.[6] That is why it is imperative for a journal site, particularly in the humanities and social sciences, to be referenced in the databases of key search engines and to be strategically positioned in search engine results listings. Referencing and strategic positioning are indispensable for achieving the objectives of a distribution site.

Referencing is a circular and developing process; registration is never definitive and has to be adapted to meet new visibility needs. Effective referencing requires knowledge of search engines, indexing methods and the editorial policies of directories. In the case of search engines, crawlers automatically index a site regardless of whether it has submitted a request to be incorporated in the database. Knowledge of indexing methods allows Web sites to be optimized for various search engines, such as AltaVista, Google, Northern Light, America Online Search, FAST Search and Netscape Search. In the case of directories that use human "editors" to classify Internet resources and apply a classification policy, the process is different. The human editors decide, based on criteria such as content quality, information organization, content accessibility, update frequency and information longevity, whether or not to incorporate a site in the directory. Examples of such search engines are Yahoo!, La Toile du Québec, Lycos, LookSmart, Francité, Nomade and Open Directory. Provided a site can be appropriately listed almost anywhere, it would benefit from channeling its referencing resources to the major search engines most likely to attract many users. Obviously, this should not be at the expense of appearing in specialized search engines intended for more targeted users that fit the site's mission.

Strategic positioning is optimal if the dissemination platform appears in the first twenty or thirty results of a search engine query. When registering in directories, a site should take care to be incorporated in the database. To that end, the site needs to identify the criteria used by that directory in order to adapt to them. Staff responsible for referencing also need to scrutinize the keywords, categories and headings under which the site will be listed and searched. It is crucial for ensuring the best possible positioning to use vocabulary corresponding to the vocabulary users are likely to use in a search. Regular monitoring is needed to ensure site positioning in search engines, because the algorithms for calculating the relevance of the search engines change, and new sites appear in the same niches.

VISIBILITY IN DATABASES

In addition to visibility in search engines and directories, humanities and social sciences journals can be referenced by large databases that specialize in scanning the content of journals.

That is not a new practice in itself, and journal editors have been active for some time in ensuring that their print publications are indexed by several databases, focussing on the most prestigious and best-positioned databases in their field. Databases tend to be selective and choose journals on the basis of their quality and influence in the field of specialization. Language also plays a role. Very few databases index great quantities of French material — moreover, there are very few databases in the Francophone world — and English-language databases generally set stringent criteria concerning the language of communication.

Nevertheless, the objective is to increase the visibility of journals in the humanities and social sciences, both in Francophone and Anglophone databases and indexes.

To facilitate better reception in Anglophone databases, it is advisable to produce consistent abstracts in English for all articles. More formalized than the abstracts generally prepared, such abstracts have the advantage of improving the conditions for selection in large databases, facilitating on-site consultation, increasing the dissemination of articles among non-Francophone users and enriching the metadata for searching abstracts. Abstracts may include particular headings, such as 1) contextualization, 2) objectives, 3) methodology, 4) results and 5) findings.[7] In this way, abstracts constitute a highly useful scientific communication tool, particularly for indexing in databases.

PROMOTIONAL TOOLS TO INCREASE SITE VISIBILITY

Greater visibility, increased numbers of users and the impact of the distribution site in the international scientific community depend on the site's capacity to make that site an intellectually dynamic space that nourishes, welcomes and stimulates activities to attract the national and international research community. The following four features are instrumental in achieving these goals.

1. Site *communication* and *interaction* with researchers and students are factors of primary importance. Activities to develop this feature include providing an infrastructure to gather comments from users, facilitating a discussion list or forum that promotes the exchange of scientific ideas, organizing on-site

events (e.g., debates, virtual conferences, first-run launches, online interviews), circulating an e-mail newsletter, providing a current events section and offering free access to as much information as possible.

2. The relevance, quality and value of the *services* speak for themselves. Their design and implementation must contribute to those features. We have already discussed selective information dissemination and reader assistance; to that could be added a personalized interface with the distribution site and a list of relevant links to sites of related interest.

3. As a player in scientific communication, the site has every interest in being part of an international network, hence the strategic interest in *partnerships*. Some simple practices contribute to that: establishing a network of journal sites (see Chapter 5); conducting reciprocal link exchanges with sites that pursue the same mission; implementing an advertising banner exchange (a form of publicity that is targeted and completely free of charge); and sponsoring smaller sites that deal with the same subjects to help them gain visibility while facilitating exchanges of services.

4. Advertising, intended to make the services of the site known, should essentially follow an *information* approach: mass e-mailings to discussion groups and distribution lists to publicize the site, prospecting through mass e-mailings to potential users, a press release announcing the site to journalists likely to relay the information in their respective traditional

and new media publications and a virtual conference with journalists in the field and important players in the community to promote the services offered by the distribution site.

Whatever methods or arrangements are used, promotion of a journal dissemination site relies on dynamism and creativity. Ultimately, the most crucial promotional factors for a journal site are quality information, document richness, services, proper maintenance and frequent updates.

ACCESS: FREE OR FOR A FEE?

The question of whether journals should provide free access plays a central role not only for Web dissemination, but also in the journals' ability to survive. Among the journals that have been publishing digital versions for a number of years, many offer free access to back issues dating back eighteen months or more. A journal that distributes its entire collection on the Web, by means of retrospective digitization, is also likely to do so free of charge. The question of filtered access is all the more acute for the publication of current journal issues.

Recently, an open letter signed by thousands of researchers (a Public Library of Science initiative), recommended free distribution of articles six months after the date of initial appearance in a journal. The particularly short period may be interpreted as an indirect form of free access. It sparked public debate on the conditions for access to journals and, in particular, free access.[8] In its wake, the *Budapest Open Access Initiative*[9] was launched in February 2002. The introductory statement points out that "Achieving open access will require new cost recovery models and financing mechanisms," a goal that is attainable. Sources of funding other than commercializing journals have to be explored, and a number of alternatives exist.

One of the major questions any journal publisher (in the broad sense of the term) must answer, at least temporarily, is the method of funding digital publishing and dissemination. Ultimately, this question concerns the organizational and financial livelihood of the journal, whatever its source of support. The equation can be stated relatively simply but is nonetheless problematic. Publishing quality journals requires resources for any publisher, whether a university press, library, learned society, commercial publisher or institute. Free dissemination of the journal would then require a source of funding for those resources and often also reliance on volunteer contributions.[10] As most journals are already experiencing difficulties, providing free access has every likelihood of exacerbating their financial instability, possibly even forcing them to "close shop" entirely. The only possibility for a journal in this situation to provide free access is to receive an appropriate sum of *ad hoc* funding.[11] When trying to envision this type of funding for many journals, we quickly come to the conclusion that the public funding network for scientific research, publication and dissemination would need to undergo an all-out restructuring. That reorganization merits careful attention; it would be unfortunate to skirt the issue and avoid debate.

CURRENT SOURCES OF REVENUE

Overall, public authorities play a major role in research development and dissemination. They support research, facilitate the dissemination of research results and contribute to the preservation and dissemination of documentation produced with libraries. In Canada and France, in particular, journal publishing and distribution have fairly similar funding mechanisms. To identify the components of an economic model that could sustainably support the digital dissemination of journals, we must turn our attention to the funding structure currently in place for the production and distribution of journals. As it stands, this environment is focussed on the print version.

A review of the financial statements of twenty-seven Quebec journals that receive Canadian federal public funding is revealing.[12] Although we cannot claim that the situation is identical in France, our discussions with journal editors and publishers suggest that it is similar, particularly when considering independent journals and journals published or distributed by publishers that are not part of large commercial groups. Thus the orders of magnitude derived from the data reflect the more general reality in Canada, a reality similar to the situation in France.

The funding structure essentially has two main sources. Public grants constitute the largest source of funding and account for nearly 47% of revenue.[13] The second major source (nearly 41%) is revenue from journal sales; of that revenue, single-copy sales are fairly marginal, while subscriptions make up 95% of sales revenue. Other sources of revenue are fairly disparate and ultimately not very numerous; among them are advertising, university contributions and other grants.

The revenue situation, in addition, of course, to the overall spending structure, serves as a basis for discussing free access, whether in principle or in application.

FREE ACCESS

Free dissemination of information, and scholarly documentation in particular, *promotes accessibility* to the extent that consultation statistics are clearly higher than if access were provided for a fee.[14]

Free service is a determining factor for users. In that sense, "free" access for users through institutional subscriptions could produce comparable results, because the entry barrier is lowered for researchers and readers who use the service online through a subscription by the institution they are associated with. Those services, referred to as *proxy services*, have done much to expand access that is largely perceived as free by front-line users. Although such arrangements ultimately rely on subscription payments from the institutions, it is safe to say that they have *de facto* contributed to and promoted free access.

Apart from the delicate funding structure that allows journals to survive as institutions of scientific communication, free access appears at first glance to be an option of choice. However, price free is not cost free! Who should ultimately bear the production and online distribution costs of free service?[15] Recently, when arXiv, the preprint server developed by Paul Ginsparg (a real pioneer in the free dissemination of research) moved to Cornell University, it was pointed out that the service required an annual injection of US$300,000 in funding.[16] That should be no surprise. In dealing with free access, whether to a preprint server or a journal distribution site, the question arises: What makes free access possible? Three hypothetical answers to that question have been proposed. We will come back to them shortly.

Another aspect of free access concerns the "market credo." It is argued that charging fees gives an indication of the real value readers attach to a journal as well as sanctioning the match between the journal's editorial policy and the readers' expectations. That may have some credibility and justify involvement in the commercial network. Yet, ultimately, small, non-profit publishers are much more driven by having to earn revenue to support the costs associated with journal production. That reasoning is correct, in a broad sense, for non-profit publishers, but not, of course, for large oligopolistic commercial groups. Journals usually do not opt for commercial practices because they identify with the ethical implications, but rather because they are forced to in order to survive. As long as their survival is not at risk, providing free access would be an attractive option.

Under current conditions (still referring to the same sample), 41% of revenue comes from sales of the print version. Physical production and shipment of the print version represent approximately 25% of journal expenditure. The major data in the equation show that free access would adversely affect the financial situation of journals, even if the journal was to achieve a substantial decrease in expenditure by eliminating the print version.

Two further considerations come into play. First, it is hard to envisage eliminating the print version of journals already using that media, at least in the near future. Second, as already emphasized, the technical production of a journal requires considerable resources for production and preservation, distribution on the Web in various formats, the production of metadata, digital preservation, the application of permanent referencing standards, access and consulting services, a Web distribution strategy, a selective information dissemination service and intellectual and physical infrastructure components, such as technology monitoring, hosting and server security. All of those factors exacerbate the problem of free access.

FREE ACCESS THROUGH RADICAL EXPENDITURE REDUCTION

Three hypothetical solutions for funding free access have been proposed. The first is based on the belief that the introduction of digital publishing allows journal production costs to be reduced. The reduction in the cost structure may target the costs of technical production of the journal as well as the costs associated with the journal's organizational existence.

In terms of technical production costs, introducing digital publishing is far from discovering buried treasure. Certainly, eliminating the print version frees some room in the budget, but still remains fairly far from some estimates, because the costs associated with the print version only represent approximately 25% of overall journal expenditure. Apart from the fact that eliminating the print version entirely may not be desirable in the immediate future, digital publishing can only generate major savings if editing and standardization work is also eliminated (as mentioned). In addition, very little is said about the various services that flow from consistent, effective distribution. In other words, there is no mystery; it is no use to dangle the hope of buried treasure if the intent is technical production of journals with professionalism under proper conditions, compliant with standards compatible with the triple mission of production, dissemination and preservation. The same functions will not necessarily require the same resources as current conditions in print publishing do; however, experience tends to show that, if we are to continue providing professional, quality services similar to those in the print environment, savings will not be sufficiently substantial to support free journal distribution.

With respect to journal organization, the ability to reduce the cost structure has also proven limited. Costs for editorial work and for journal administration are difficult to reduce in that they have already been kept to a minimum, relying heavily on volunteer contributions either through service loans or volunteer work. Current costs only represent a part of the resources necessary for their survival. With the introduction of digital publishing, prospects for re-engineering the editorial process, peer review,[17] proofing, copy preparation and so on are particularly exciting. On the whole, we can computerize and digitize the entire process from the submission of a manuscript to the publication of the journal.[18] Digitization is implemented in phases and requires an initial investment that is not negligible. Over time it will be a source of sustainable savings yet the exact cost reductions can only be approximated. In the foreseeable future, unless jeopardizing the physical and organizational survival of journals or asking the associated institutions, whatever they may be, to significantly increase their contribution in services and in kind, free access can only rely on reducing the costs of editorial structures.

FREE ACCESS THROUGH PAYMENT OF PUBLICATION COSTS BY AUTHORS

The second option would be to provide free access to online journals by collecting a financial contribution from authors. In this scenario, authors would pay a fee to be published. Thus, it is neither the journal nor the journal users, but the individual published, the author, who provides free access.

That practice, which already exists in the print publishing environment, is more common in the natural sciences and technology-based disciplines.[19] Occasionally, authors are charged for publishing in the social sciences and humanities; however, the contribution remains modest and most often appears in the form of a processing fee (e.g., for processing images). Let us look at production conditions in respect to the feasibility of charging authors. First, although the practice is current in some disciplines, it is fairly exceptional in the humanities and social sciences. Financial contributions by the authors to be published are not customary, particularly in journals published by non-profit organizations.

Introducing a publishing fee for authors would risk meeting major resistance. It assumes authors have the financial capability to meet journals' expectations. Unless research funding is to receive a substantial boost, that assumption is unwarranted. University professors do not have a budget at their disposal to pay such fees unless they receive research grants. The success rate of research proposals submitted to major grant competitions in the humanities and social sciences in Canada is approximately 35%. The lucky 35% of grant recipients thus represent only a fraction of the whole body of professors, which means that author publishing fees would introduce a form of segregation among authors based not only on the quality of the work to be published but also on the ability to pay for publication.

The only way to render the practice almost painless would be to reduce the fees to such a low level that they would be symbolic in nature. What is the threshold for a symbolic fee? C$50 or C$100 (35€ to 70€) per article? Whatever the threshold may be, it is obvious that the downside of introducing such a symbolic fee, in terms of author relations, image and so on, is not worth it, given the modest amounts collected. (We are talking about a total amount of C$900 or C$1,800 (600€ to 1200€) a year for an eighteen-article journal.) The option of charging authors is therefore a questionable solution.

FREE ACCESS THROUGH RECOGNITION OF DIGITAL JOURNALS AS A PUBLIC ASSET

With this third option, free access is not linked to the supposed drop in journal costs nor to "disbursements" by the authors published, but to the journal's status as a vehicle for disseminating research results. Identifying journals as a public asset radically changes how publications and their production and dissemination are viewed.

A whole series of goods and services have been identified as public assets, to which access is now largely facilitated or without restriction. To provide access, funding is generally found through public channels. An asset is not public by definition, nor is it excluded from that designation outright. Things evolve and are differentiated depending on the society, which may change at will. For example, a society may invest significantly in social housing for some time and then withdraw funding for that cause at a later date. In the same vein, access to books and reference material may be promoted through a library network project, only to be relegated to the pile of low-priority issues some time down the road.

Although most stages of research do not take place in a purely commercial setting, because they are supported directly or indirectly by public funding, the dissemination of research results through journals is subject to that setting. We could just as easily think that, due to the interest and usefulness of the content they publish and the visibility they provide for the national research system, journals should benefit from public asset status and the broadest possible access. It is true that public grants support approximately half of journal budgets. Yet, even at that, journals are obliged to resort to marketing networks for a significant share of their revenue — a *sine qua non* condition for their survival. We can envision journals and the content they disseminate being recognized as a public asset, because they are the natural extension of a chain of key links supported by public authorities. That is why it is important to ensure the most effective possible distribution.

That option has gone beyond speculation, materializing in January 2000 with the announcement by the National Research Council of Canada that the digital versions of their fourteen journals could be freely distributed to users with an IP address ending with <.ca>. The project was made possible by the commitment of a federal government agency (Public Works and Government Services Canada) to cover the loss of revenue resulting from that measure in Canada. The government contribution was based not on the replacement of lost revenue, but on the cost-recovery principle. Printing and distribution costs were thus subtracted from total direct costs to determine the first-copy cost, in whatever format. From that were deducted journal organization and peer review costs, i.e., costs associated with the journal's existence as an institution and the manuscript selection process. The result of the calculations gave an estimate of the production costs of the digital version. The public contribution was calculated on that basis.[20]

Another avenue can be envisioned that draws on the contribution of university libraries to ensure free access for users. Libraries can already be said to "subsidize" journals through institutional subscriptions or, based on a practice now current for digital journals, through user licences negotiated institutionally, often as a consortium. A further step would be for a library to subsidize journal production at the source, and journals then be disseminated free of charge by a journal site. That type of partnership, formalized in multi-year agreements, would provide funding for journals while stabilizing library subscription costs. The Scholarly Publishing and Academic Resources Coalition (SPARC) recommends similar models. To pursue this path, the libraries and their respective acquisition rates of national journals must now be examined to estimate whether they would be in a position to provide sufficient funding for journals. That does not appear to be the case for a small society, such as Quebec, and is far from certain for Canada as a whole; however, it would probably be the case for France. In any event, the net contribution from that source would undoubtedly be supplemented by government grants and other sources of funding. The structure could thus offer a viable economic model.

Whatever the model, relying on the contribution of a "public sponsor" would undoubtedly apply best to the situation of journals and their complementary objectives of quality production, wider distribution and organizational stability. The major difficulties are acquiring the recognition of public asset status for the online distribution of journals, followed by capturing the "interest" of a public agency to support that public asset financially. This scenario certainly offers the best conditions for upgrading digital publication and evading all-out commercialization for which journal publishers show little attachment.

EFFECT OF FREE ACCESS TO THE DIGITAL VERSION ON THE PRINT VERSION

What impact may free dissemination of the digital version have on the future of the print version? All our consultation with Quebec journals and surveys of the practices of major humanities and social sciences journals (that are print publications) lead to the conclusion that the print version can be expected to remain for an as yet undetermined period of time. We can take that expectation as a given and question the interaction between the two media, particularly if different economic conditions for access apply (i.e., free access on the Web and subscriptions for the print version).

Print versions still meet real needs of readers in addition to equally real demands for the consultation and distribution of journals; it is hard to imagine that exclusively online distribution could stamp out inclinations and preferences for the print version. The results of some experiments in this area also appear to go in that direction.[21] On the other hand, we can anticipate that free online access would have a negative effect on the level of subscriptions to the print version. But to what extent? Even assuming current subscriptions are reduced by half, the direct costs of paper, printing and shipment for the print version – revamped with more effective production processes (digital printing) – should be covered by subscription revenue. We might also ask whether the demand for the print version is large enough to justify maintaining it. Ultimately, taking all the factors into account and still in approximate terms – a decrease by half in sales revenue and corresponding drop in print runs, with technical modifications, where necessary – the cost of the print version could very likely match sales revenue (i.e., subscription sales and single-copy sales). In the medium to long term, depending on the particular situation of the journal, we can also envision a single issue per year of the print journal.

All those issues are open and leave room for various assessments, given the level of uncertainty about the reaction of journal editors, subscribers and readers and the possibility of a public sponsor. However, we cannot sidestep a significant concern that may arise for any of these options. Free access reduces independent journal revenue, increasing a journal's dependence on public authorities for their survival beyond production of the print version *stricto sensu*. A journal's position can then be considered extremely vulnerable to public policy, from a financial and potentially an editorial standpoint, which creates a need for strong assurances of the determination of public authorities to provide sustainable, satisfactory funding; otherwise, journals would rightly be reluctant to accept a bet that could be a very high risk.

FILTERED ACCESS FOR A FEE

If free access is impossible, because the necessary factors are not all in place, we have to resort to filtered access for a fee at a distribution site. The situation is not atypical, because one form of filtered access or another is in effect at key journal distribution sites, including non-profit sites, such as MUSE,[22] HighWire[23] and BioOne,[24] to name only a few. As a result, a journal marketing method and subscription management system have to be designed. Charging a fee for consulting digital journals is common on the Web, with some procedures emerging as more prevalent, although we are still far from reaching consensus.

The recurrent issue of market segmentation in discussion and practice finds applications that, at times, head in diametrically opposite directions. Recent developments instead suggest that simplicity should be preferred in any method of revenue collection.

Starting from the distinction in production costs between the costs associated with production of the first copy (i.e., fixed costs) and the costs for additional copies (i.e., variable costs), Colin Day[25] proposed segmenting readers by introducing price discrimination mechanisms. If one segment of readers were to pay a subscription price that would cover the much higher first-copy costs and the other segment were to pay a price that reflects the marginal costs, the very nature of the services would either vary significantly or be aimed at clearly distinguished users. On one hand, upgraded and diversified services have to be delivered at the highest cost; on the other hand, basic services have to be offered at a reduced rate. We can also envision libraries and institutions, which act as relays in journal distribution to many readers, having to pay a price that covers first-copy costs, with individuals paying a subscription rate that covers marginal costs.

Referring to the specific situation of learned societies, Varian[26] recommended differential pricing based on the observation that there are two categories of readers of journals produced by learned societies, the society's members and non-members. The journal is delivered to members as a benefit of their annual dues. Journal consultation at the library would then apply largely to non-members. In the case of print publications, the convenience of owning a copy of the journal rather than going to the library makes sense. In the case of digital publications, however, distance consulting through the library may result in a drop in membership and "automatic" subscriptions, thus a major loss of revenue. To meet the challenge, journals must make their individual subscription more attractive by differentiating the services offered. Such a campaign would increase the marginal benefits for an individual subscription over the benefits of in-library consultation. From there, it would then be possible to restrict access to the institutional library subscriptions to university campus networks, through licenses, blocking access to communication from off-campus terminals. Yet restrictions of that kind are essentially defensive moves, no longer very relevant with *proxy* servers. In a more positive vein, it is also possible to consider the potential for enriching and diversifying the service offered to individual subscribers only, such as hypertext links, a higher-performance search engine, selective dissemination of information service and the privilege of first access to new articles for a specific period of time.

That reasoning clearly demonstrates the issue of individual subscriptions for the financial survival of a number of journals. Yet, at the same time, it underlines the difficulty of finding an appropriate solution. It is difficult to justify, for example, libraries paying more and receiving less service. Moreover, it is fairly understandable why many university presses have decided to sacrifice individual subscriptions and keep only institutional subscriptions.

Segmentation is also often based on the services offered. Subscription price scales are thus often determined by the versions used. Various methods in use reflect different alignments between the print and digital versions. In the following section, we briefly outline the practices of some university presses, in order to clarify the issue.

DIVERSE REVENUE COLLECTION METHODS

This review of practices in recent years shares the viewpoint of university presses and learned societies, which are first and foremost concerned with cost recovery, not charging oligopolistic prices, as in the case of commercial publishers. Presenting journal publishers' revenue collection methods can be particularly complex, if we intend to produce an inventory of practices. For the purposes of this discussion, we must content ourselves with identifying the key revenue collection methods practised by university presses in recent years.

- *Access to the digital version is free with a subscription to the print version of the journal.* The cost of the digital version, in this case, is financed by the revenue generated from conventional sales of the journal.[27] That raises three different concerns. First, the model is only possible if the journal's revenue significantly exceeds production costs for the print version and the digital version incurs low or marginal expenditure. Second, this model discourages access to just the digital version of the journal, which limits the choice of potential readers. Third, in the short term, this formula avoids addressing the problem of negotiating prices and services with libraries, because the financial data remain essentially unchanged, but are now coupled with additional service. We might add that, should this practice continue, it would result in obscuring the costs of digital publication, sending the message that it costs nothing.

- *The print version is not sold alone, but bundled with the digital version; however, the digital version is offered alone for a slightly lower subscription price.*[28] In this case, the costs recovered from producing the journal are almost equally distributed between the two media. In addition, there is a fairly good incentive to maintain a subscription to both media, due to the low variance in subscription prices. This practice may favour varying the timing of delivery of the two media. Although it may be seen as a drawback, the presumed advantage is that, because the digital version hardly marginalizes the print version in terms of price, the journal's survival is not threatened by a drop in print subscription revenue. For journals that have low profit margins or anticipate rigidity on the part of their readership, it is certainly a safe option.

- *The pricing scale varies significantly based on the status of the subscriber or type of services. The scale reflects the relative cost*

of production of the first copy and additional "copies" for different media or is used to promote a change in subscriber behaviour.
In those cases, we note wider price variances. MIT
Press, for example, has a two-tier pricing structure.
The first tier reflects the production cost for the first
copy (i.e., all fixed costs including editing work and
selection) and is intended for institutions (US$160
to US$205). The second tier, intended for individu-
als, reflects the marginal production cost (US$40 to
US$50).[29] The variance in prices may also be part of
a strategy to induce subscribers to retain only the
digital version of the journal, thereby facilitating
phasing out the print version. Management of the
Journal of Biological Chemistry decided to discourage the
simultaneous purchase of the print (US$1750) and
digital (US$1300) versions of the journal by offering
no price discount for those buying both versions.[30]

• *Subscription prices to the different media are based on the production costs for each medium, and the pricing scale allows separate subscriptions to either of the media or the package.* This
model was proposed by Project MUSE leader, Johns
Hopkins University Press.[31] Essentially, only institutional subscriptions were offered. The model is
both flexible and simple and provides fairly good
visibility for what we could agree to call the over-
head costs of the journal and the costs associated
with the different versions. Individuals' access to
journals is tied to the subscription of the institution they are associated with; users are filtered by
IP address. Losses from individual subscriptions are
meant to be offset by the number of institutional

subscriptions, which can foreseeably be increased with careful upgrading of the package deals offered. This option is only possible, however, when a sufficient number of journals can be offered as a package deal, thus implying a fairly large volume operation. The package-deal option constitutes a definite advantage for certain publishing and distribution structures.[32]

Many of the above practices are also prevalent among the commercial publishing oligopolies that dominate journal sales, the figurehead being Elsevier's ScienceDirect. In that case, however, regardless of the practice, a different logic applies – the logic of prices charged to collect the largest possible guaranteed revenue. The free-market credo of unbridled profit orientation predominates and has led, in turn, to the establishment of library consortiums that hope to mitigate the power relationship and to improve conditions for negotiation.

SUBSCRIPTION MANAGEMENT

This review of the various revenue collection methods has highlighted the underlying logic for setting prices, independent of their levels, and how institutional and individual readerships are treated.[33] Added to that is the need to design a model for the functional relationships between subscriptions to the print and digital versions of a journal. Some considerations deal specifically with the operational infrastructure.

The starting point is to give a distribution site a front-line role in subscription management of print and digital versions.

Journals currently manage their subscriptions themselves or have them managed by a publisher or a specialized commercial or university subscription agency. There is thus a large number of subscription management sites for the total number of journals. A distribution site for several journals would simply act as the main entry point for subscriptions; there would only be one database, one protocol and, potentially, one pricing framework. However, that involves an all-out change in the way journals currently operate. In addition, where individual and institutional subscriptions are maintained, two methods of filtering access would be required, i.e., access by IP address for institutions and access by password for individuals. Individual subscriptions also require an infrastructure for online payment, while institutions generally transfer payments differently. As far as that goes, the effect institutional subscriptions to the online version may have on individual subscriptions has to be considered, because access to journals through institutional subscriptions is subject to increasingly fewer restrictions; with a mandatory server, no one needs to consult a journal from campus, for example. There will certainly be a shift, perhaps even a reduction, in the number of individual subscribers to the digital version.[34] In this context, should a filtering system for the digital version be established for individuals? In other words, should individual subscriptions be offered only for the print version?

The previous comments support the option where the distribution site provides filtered access only for institutional subscriptions. That formula is fairly widespread for three main reasons. First, it is user-friendly; second, it allows researchers to consult a large number of journals on a one-time basis (i.e., that does not warrant a subscription); and, third, it is fairly simple to implement technically and organizationally and is increasingly aligned with practices in the research community. After free access, it is the option that most serves to promote consulting articles online. Individuals would be in a position to subscribe to the print version by contacting the journal or its agent or by putting pressure, where necessary, on their institution to subscribe to the journal.

That said, there is nothing to prevent a dissemination platform, which would only handle institutional subscriptions, from giving preference to package sales, i.e., sales of the whole collection of journals distributed. This model, which has been used by oligopolistic commercial groups as a sales method, is used increasingly frequently by libraries that find it in their best interest as buyers to form consortiums. In those cases, however, bundling is not being used to achieve the same aims as oligopolies. Nevertheless, an "all or nothing" approach may be considered rigid by specialized institutional buyers, such as research centres and universities in other countries that are not interested in a subscription to the whole collection. For that reason, to achieve the desired flexibility, arrangements need to be made right from the start so that the package deal is combined with the offer to subscribe to select journals. In any case, transactions between the platform and subscribing institutions (whether as single institutions or as part of a consortium) should be as simple as possible. Finally, we consider the package purchasing model advantageous for individual journals because it increases the dissemination of many journals that become accessible to libraries and their clients who had not subscribed to them previously.

SUBSCRIPTIONS, E-COMMERCE AND
REVENUE EXPECTATIONS

Providing access for a fee (for individual and institutional subscriptions) involves a journal dissemination system that includes subscription management, access control and access statistics, "immediate delivery" of articles, multiple formats of the same article, management of the title catalogue and pricing scale and account management (including payment deposits and revenue allocation). This infrastructure requires the resources for its implementation, followed by resources for its sustained operation. The cost of those resources depends on the range of options offered for individual and institutional subscriptions to the digital and print versions. For example, sites wishing to offer only institutional subscriptions require less resources than others; the volume of transactions decreases and the system for filtering access is simpler. Greater use of computer processes should allow greater optimization of the resources mobilized for management.

It is fairly difficult to anticipate the reactions of users to the emergence of dissemination platforms with full-text access, subject to subscription. In principle, the service should elicit a certain fad interest; however, at the same time, some slowness to adopt new developments may be encountered in individual and institutional behaviour. The impressive expansion of e-commerce might ultimately be expected to carry over to all types of content, including scholarly documentation. However, failures and nosedives in a number of e-commerce sectors have done their part to tone down high hopes. Therefore, the cost-effectiveness of selling subscriptions to the digital version of journals requires detailed study.

So far, the sale of digital journals has not fulfilled revenue expectations, a trend that is likely to continue, and has not succeeded in generating sufficient funds for the transition to digital publications, a transition which most university presses and research organizations have had to finance through considerable financial support.

On the other hand, we can envision deriving new sources of revenue from digital distribution, for example, through access to new readerships that print versions cannot reach. Similarly, publishers may discover new vehicles for distribution, for example, minor players now entering the scene by joining consortiums, such as libraries in colleges, municipalities, preparatory schools and high schools. Additional revenue may also come from resellers in the form of fees per article.

While exercising caution in our expectations, we should note that libraries have confirmed their interest in digital scholarly documentation. That interest is demonstrated, in particular, by their acquisition policy. Everywhere, significant investments – perhaps insufficient, but nonetheless significant – are allocated for the equipment needed to provide quality service in the consultation of digital documentation. In a few years, the impact of that transformation on the collection of revenue from digital documentation, particularly from journals, will have to be assessed.

The evolution of digital dissemination practices for scholarly documentation, particularly journals, quickly renders current considerations and expectations obsolete. Uncertainties and problems have still not dissipated in the turbulent ongoing change experienced by the Web. They are still being raised, whatever their particular nature, but require answers and operating methods that are formulated in a dynamic perspective. In other words, the revenue collection method implemented cannot avoid being assessed against some of the issues raised in this chapter, particularly free access to research results.

IMPLEMENTING AN ECONOMIC MODEL

For many, the stumbling block in a digital dissemination system remains an economic model that could ensure its funding in a non-profit structure. In the case where the necessary conditions for free access have not all been met, this issue is all the more likely to be raised because, for many, an Internet revenue collection method conflicts with the image and fantasies associated with the Web that support free access. That said, Web transactions to access digital material are becoming more common. Moreover, established journals published by university presses and learned societies often offer filtered access to the corpus of journals disseminated digitally, involving payment of some sort.

That is neither a fantasy nor a profit obsession, but a minimum condition for ensuring the economic life of journals. As long as we have not succeeded in "re-engineering" public funding networks for research and its dissemination, journal editors and their partners have to view digital dissemination through filtered access for a fee as a necessary given.

Some reference points can now be identified. The revenue collection method should be simple and give some visibility to the real production costs for print and digital media; in that sense, modulating the price based on the service or product delivered has proven advisable, without the need to obscure it in complexity. One and the same dissemination platform may very well practice more than one pricing logic. The pricing would be based on an annual subscription, preferably for institutions. Without opting for marked variances, institutional subscription prices should reflect the status of libraries as relays in the communication of reference material to a wide readership, as well as their strategic position in the scientific communication system. In that sense, they are partners for whom package-deal journal pricing is clearly well accepted. Earlier, we identified five approaches to pricing logic in practice. Each approach has to be evaluated considering the objectives of each journal. Concurrent to those approaches, it would be hard not to develop an approach that highlights journal package deals for institutional clients.

4: TECHNICAL and ORGANIZATIONAL PROSPECTS

All the variables highlighted up to now inform our decision on the action to take and the measures to adopt. This reflection is inspired by current debate on digital publishing of scholarly corpuses and the results of experiments in the academic setting. At the same time, this reflection benefits from our experience founding and developing *Érudit*, a digital publishing and dissemination platform. The lessons learned so far serve as a starting point for outlining prospective digital publishing infrastructures. Those prospects are part of a much broader movement that is taking shape in a number of countries, including the United States, Canada and, soon, France.

THE DIGITAL CHALLENGE

Our challenge in digital publishing is to adopt information technology to serve scientific communication. A genuine transition to the digital environment affects the use of research results, project communication, research tools, the publisher's role and the components of the scientific communication chain — from authors through publishers and libraries to researchers and readers.

Digital dissemination of journals means that retrieving and consulting research documentation occurs, to a great extent, under entirely new conditions. The new environment has a major impact on how authors process and use research results. Moreover, it allows us to explore and conceptualize a new model for journals where, with or without a parallel print version, the digital version is considered first. This would constitute a considerable shift from the current model that regards digital journals merely as a display of print material that was written within the internalized constraints of print media.[1] In the new method, documents intended to report on research processes and results could thus be conceived and produced to exploit the potential of digital methods. We cannot conceal the fact that, as it stands now, digital versions of journals are by and large considered a digital translation of the print version. Although they are in a new environment, the writing and publishing practices remain fairly conventional. Moreover, a number of journals that are produced only in a digital version have taken a fairly unoriginal approach that follows accepted practices in the print environment.

There needs to be discussion on journal design and format and the resulting practices. Researchers and journals are being asked to take part in renewing the conditions for developing and transmitting research results. This discussion and the innovation it may generate depend on the awareness of journal editors and certainly on the disciplinary sectors and types of documents. Depending on the sector, the possibility of incorporating multimedia components, hypertext links and dynamic data in the body of the journal are considered more or less attractive. The effects of the introduction of digital processes will certainly not be instantaneous, and the rate of adoption is hard to predict.

ONE TRANSITION, MANY OPTIONS

How will the transition be achieved? To answer that question, a whole set of social, institutional, cultural and technical conditions have to be considered. For example, parallel production of the two media, print and digital, is a constraint that often has to be faced. In light of local and international experiences of already well-established print journals, the transition is effected by maintaining print and digital media in publication, distribution and preservation. It is up to the various players involved in journals to determine how long the process will take, if we are to ultimately see the print version disappear.

RATIONALE

That raises the issue of the intellectual and institutional rationale for adopting a digital version of a journal. We can outline two scenarios.

In one scenario, the key motivation is an increase in dissemination. That motivation also exists in the other scenario, but clearly takes precedence in this scenario. The focus on dissemination may be motivated, for example, by a vision of the value of digital methods for scientific communication, by the shortage of available resources in expertise and equipment or by the willingness to extend the transition over time by not changing how the journal is produced at least temporarily. Those considerations may lead to opting for the production of PDF files from page-layout files of journal articles, an approach that has the merit of ensuring broad dissemination at low cost.

In the second scenario, the main target is to exploit the digital version with all its new possibilities for both production and dissemination of the journal, taking action on both fronts at the same time. The introduction of digital methods in the journal publication process itself contributes to the digital version being seen as the first version[2] by authors, publishers and readers.

Between the two scenarios, a number of different operating methods can be envisioned. It is generally agreed, and we share that opinion, that the integrated XML processing chain ensures a controlled transition for both publication and dissemination processes and offers the best guarantees for document preservation. The process is "integrated" by the production of all dissemination formats, print and digital, from the same XML file. Although the integrated XML chain may be seen as an optimal scenario, other operating methods should not be dismissed. For that reason, we consider three other options, from dissemination of a digital version only – duplication of the print version – to a structured language version that maintains a parallel processing chain for the print version.

DIGITAL PUBLISHING MODELS

An overview of international practices shows a variety of choices being made in publishing journals digitally.[3] Production of the digital version of a journal goes well beyond simply making articles available online.[4] In each of the three functions – production, distribution and management – we see a plurality of practices in production formats, archiving digital material and economic models.

Although the integrated XML scenario is undoubtedly the most promising and systematic for the transition to digital publishing, other scenarios merit explanation. The various operating methods differ depending on their specific stakes and interests in digital publication.

To simplify the discussion, we have reduced the options to four, i.e., PDF (text), HTML, page layout to XML and integrated XML. We will outline the features of each of the options in relation to thirteen key indicators (see Table 5), categorized by the major digital journal publishing functions. The choice of a model should be based on the situation and the expertise, resources and equipment available.

PDF (TEXT) MODEL

The PDF (text) model involves few modifications to the majority of production processes. The journal team receives articles from authors in word processing format, most often Word. Once the editorial process, language review and standardization based on the style guide are completed, the page layout and proofing are done using page-layout software, such as QuarkXPress, to name only one. From there, a digital file or camera-ready copy is produced to print the journal issue.

In this model, the PDF files are simply produced from the page-layout application using Adobe Distiller software. Most of the time, the operation is carried out without any problem, particularly when it is done using the computer that was used for page layout, which guarantees, for example, the presence of all the font files used. The result is a PDF file that faithfully shows

FIGURE 2   PDF (Text) Model

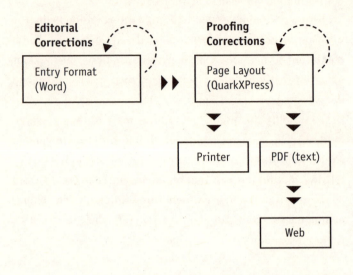

the image of the print document on screen as it will come off the press in a dissemination format that could be called a *screen print*.

This production model for a digital version merely constitutes a translation of the print version available on line. Despite the hypertext potential of the PDF format, it is more often considered for print on demand. The conditions for long-term preservation are also more difficult to guarantee, because the PDF format is proprietary. Although PDF is *de facto* a standardized format and Adobe (the company that owns the PDF format) provides ample information and technical details on the format, only Adobe can make any change whatsoever to the format, without having to consult anyone. Research underway on this issue portrays conversion of PDF to another format (image or text) as the solution for ensuring the longevity of information encoded in PDF format.[5]

In response to the importance of XML in the industry as a whole and particularly in the publishing community, Adobe has developed a module for the Acrobat suite that allows a PDF file to be converted to XML.[6] Given the nature of input files, the result of this kind of processing may essentially be a "well-formed XML document" based on page-layout features. However, a second conversion phase is required to implement the full potential of "valid" XML files, based on a DTD or schema. In addition, to the extent that a PDF file does not have any structured information, the production of metadata cannot be done automatically, limiting the search potential to full-text searches.

The PDF model is the least costly model to implement, both for the equipment that has to be purchased, which is limited to Distiller, and the expertise needed.

HTML MODEL

Most word processing and page-layout applications allow output to be produced in HTML. However, the use of those automatic functions gives disappointing results. HTML produced that way is, by definition, an overlay on the page layout prepared for the print form. It is then necessary to rework the files by deleting the unnecessary codes and modifying the automatically entered markup tags. Although that procedure is not very costly in equipment and software, it is costly in terms of the time required for arriving at a minimal dissemination format (HTML). It is often more

FIGURE 3   HTML Model

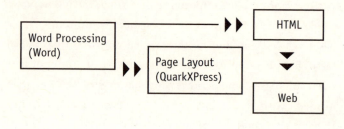

opportune to use tools that allow better control over the insertion of tags.

With its limited set of tags, HTML does not allow structured searches, and compromises often have to be made to identify the different text elements. Although tags such as "H1" and "Blockquote" give an indication of the nature of the information they contain, their unsystematic use and the absence of tags to identify the majority of elements in an article restrict the search and display potential. For the same reasons, the automatic production of metadata is impossible. Because HTML is an SGML application and its basic character set is ASCII (or UNICODE for XHTML), some longevity can be assumed. However, the use of tags specific to particular softwares and the juxtaposition of information on information display (e.g., "Center" or "Bold") do not make HTML a format of choice for long-term preservation. In fact, HTML is essentially a dissemination format. It was not designed for information production or archiving.

PDF and HTML solutions are often considered limited to meet the publication, dissemination and preservation needs of journals in digital form. However, the implementation of an integrated XML processing chain that allows the full benefit of digital potential − for the production of all distribution formats from the same XML document or the production of a real digital journal that incorporates multimedia elements − involves significant changes in operating methods. For justifiable reasons, i.e., either the expertise or the organization required, the implementation of an integrated XML processing chain may prove difficult.

PAGE LAYOUT TO XML MODEL

In the *page layout to XML* model, the journal is not confined to restrictive choices, a feature which the model attempts to exploit. Most print versions of journals are produced using page-layout software such as QuarkXPress or PageMaker. The *page layout to XML* model uses the page-layout files to produce the XML version. Conversion can be done in several ways. Some tools, such as Quark or LogicTran, allow conversion to XML. Scripts can also be used to convert page-layout files to XML. In such conversions, the text of page-layout files is first extracted to a format such as RTF. The second step is to use programs to convert the files to XML. The programming can be done in PERL, Omnimark, JAVA or VB, for example. An XML editor is often useful for fine markup, where the complexity, combined with a very large number of possibilities, does not justify designing and developing an automatic conversion program. Once the XML format of the articles has been created, the dissemination formats are generated automatically in HTML,

XHTML or any other dissemination format, using XSL style sheets, for example.

The primary disadvantage of this model is its dependence on a proprietary format, such as Quark. We must admit, however, that this disadvantage is relative, because the use of proprietary software is current and well-established for the page layout of journals. In this scenario, where programming for conversion is done outside the page-layout application, the input format is a "text" format, produced from a proprietary format. The second disadvantage of this model is that it makes the digital version the product of a process designed for the print version. The production chain is print-based; once the journal is ready for printing, a file is sent to the printer and used for conversion to XML operations at the same time. Nonetheless, given the prevalence of the print version and the expertise and proven operating methods of journal teams, the page layout to XML option is certainly one to consider and easier to implement than an integrated XML processing chain. A third irritant in this chain is the structure of the page-layout files themselves. A number of page-layout tools operate in such a way that the document is not a "ribbon" of text that unrolls page by page. On the contrary, different text elements constitute a number of "boxes" independent of one another. For example, footnotes are in a separate "box" of text, and the reference marks are not linked to the footnotes themselves. Similarly, tables and figures, for example, are not related to text. Features unique to the design of documents intended for printing create problems for extracting the full, continuous text, problems for which various solutions have to be found.

FIGURE 4   Page Layout to XML Model

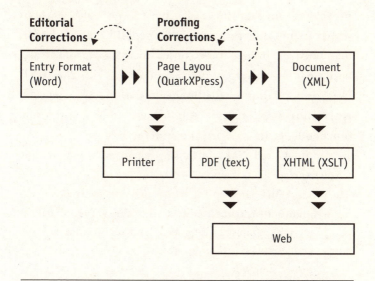

Because the majority of authors write for print distribution, this model offers the journal the exact same advantages as the integrated XML model. The variables that affect search potential, automatic metadata generation and longevity indicators are similar for the two models.

The resources, document analysis expertise and programming skills needed for implementing this model are similar to those needed for the integrated XML model.

INTEGRATED XML MODEL

The integrated XML processing model is based on principles that meet the objectives for publishing scholarly journals in a digital environment.

This processing chain uses standardized formats for information display and tries to use standardized processing mechanisms as often as possible, without jeopardizing effectiveness and flexibility as a whole. Standardized information coding formats are necessary, because they offer some independence in the tools used and guarantee the longevity of the information. No standardized tools are proposed, thus maintaining independence in software and systems.

An integrated XML processing chain may include the following stages: [7]

1. manuscripts submitted by authors are standardized for bibliographical references and so on, and language review is done in a word-processing format, most often Word;

2. a Word version with style sheet is prepared, applying styles to the text that identify the different text elements, such as title, author and heads;

3. the articles are then converted to XML;

4. from XML, a first page-layout version is prepared to produce a set of proofs;

5. corrections are entered directly in XML; and

6. once the copy has been approved by the journal, the different versions are produced for distribution, for both the print and digital versions of the journal.

FIGURE 5   Integrated XML Model

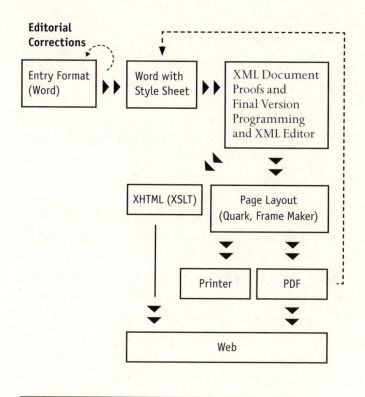

A journal production method should allow the multimedia and interactive nature of digital information to be exploited. The integrated XML model facilitates that exploitation. Digital documents stimulate changes in operating methods, distribution and use of research results, while opening up new perspectives for production. In addition, they allow the integration of different types of information, from text, 3D animation, image, video and sound to virtual reality.

The expertise required for the integrated XML chain is diversified, which makes it necessary to involve professionals. Staff costs for the implementation and operation of the model are relatively high.

OVERALL COMMENTS

The key points in the previous section are restated in Table 5. The four models schematically portrayed are described in the table by indicators for production, dissemination, preservation and resources.

The *production* process for a print publication is carried out in an environment where the information dissemination media is known and controlled. In the digital world, however, there are a variety of dissemination formats and, most important, those formats have a wide variety of technical qualities and features. The integrated XML processing chain meets the need for flexibility in digital or print information production, in order to service the greatest number of current and future dissemination media and information consultation experiences.

Documents structured along XML standards contain information on their logical and semantic structure. This could include information on different sections of the document and their heads or particular parts of the text, such as geographic locations and people's names. That structure makes them rich documents, because it is easy to translate structure information into interpretation information. Converting information such as "this is a first-level section head" into an instruction such as "this must be printed in 14 pt. boldface type" can be done easily and, most important, automatically. The use of structured documents is the only way to follow the "one source, many products" principle in a digital context.

The "page layout to XML" and "integrated XML" models allow a structured document to be created for each article that is sufficiently rich to permit all the planned uses. The economic interest of these models lies in automating information processing in various ways. Creation of the information always remains an intellectual process. Yet, once the information is created, its transformation into different management and dissemination products should always occur automatically.

Those observations lead to the conclusion that the commitment to digital publishing requires the specification of a series of parameters and a process. Of course, there are technical considerations; however, the observations also imply setting priorities for the workflow and the organizational structure. Defining a publishing standard assumes that the corpus or nature of the document to be published has been considered. A journal – its mission, intended readership, "useful life" and positioning in the knowledge development and accumulation process – calls for quality processing, flexibility of use and guaranteed preservation.

Evidently there are technical consequences; most of all, however, the main organizational challenge is to implement a processing chain in structured language that is fundamentally based on digital processing from the start of the production process and can generate various dissemination formats. The other options involve prior processing for the print version, followed by digital processing. Digital processing could involve PDF text processing or processing for the Web with HTML or a

TABLE 5   Indicators for Various Models

	PDF (text)	HTML	Page Layout to XML	Integrated XML
Production				
• process (electronic version first [1] or derived from print version [2]	2	2	2	1
• output format (non-proprietary [1] or proprietary [2])	2	1 and 2	1	1
• software (non-proprietary [1] or proprietary [2])	2	1 and 2	1 and 2	1 and 2
• metadata (automatic [1] or manual [2] production)	2	2	1 and 2	1
Distribution				
• format (automatic generation of several formats)	More or less	No	Yes	Yes
• search (structured [1] or full text [2])	Particularly 2	2	1 and 2	1 and 2
• data reuse (easy or difficult)	Difficult	Difficult	Easy	Easy
Preservation				
• indicator of longevity	Low	Average	High	High
Resources				
• production cost	Low	Average	High	High
• equipment cost	Low	Low	Average	Average
• expertise	Low	Average	High	High

structured language after processing with page-layout software, hence the name "page layout to XML" model.

Those choices determine the production and preservation formats, to a large extent, as well as the software with which the work is done. In general, it is more appropriate, less restrictive and more economical to use open formats and open-source or free-of-charge software, at least for the educational environment, which is why the distinction is made between the proprietary and non-proprietary nature of software and formats.

Similarly, in the best-case scenario, the method of producing metadata – that describe the attributes and content of articles and are used for information retrieval, management, description, access and preservation – should automatically extract the information from the articles themselves. If this method is not implemented, a time-consuming and demanding manual procedure is necessary.

A series of questions on *dissemination* allow us to qualify the four options selected. Depending on the production format used as a master format, different formats could potentially be produced. For some master formats, such as PDF and HTML, that possibility rarely exists, while, at the other end of the spectrum, rich formats, such as XML, allow a number of formats to be produced in a given period.

Format also has an impact on *search* capabilities and the conditions for improving the effectiveness of results by eliminating both noise and silence. In this area, the potential for conducting a structured search is an advantage over full-text search alone. It should be noted, by the way, that a PDF image does not allow either a full-text or structured search.

Document *dissemination* needs evolve and diversify over time. The richer the production format, the wider the range of possibilities, which is an advantage. For the purposes of comparison, we have used a dichotomous variable for data reuse, i.e., whether it is easy or difficult.

As long as reliable document archiving procedures have not been publicly established, journal publishers bear the responsibility for ensuring the *longevity* of the digital version of the journal. In a more modest way, in digital dissemination, journal publishers have to ensure the continuity and reliability of online service and article availability, which necessarily requires document preservation, despite and beyond the technological obsolescence of both software and media. It is therefore of primary importance to consider the longevity indicators of the various formats.

The *resources* that have to be mobilized vary according to the processing chain selected. Production costs are first and foremost dependent on the expertise needed to carry out the project, while equipment costs (including hardware and software) are relatively similar; hardware is comparable and software accounts for a relatively small part of overall costs (see Table 5).

INFRASTRUCTURE DESIGN AND DEVELOPMENT

How should we consider and define the role of publishing in the context of an information society and the implementation of digital publishing in scientific communication? The issue is not so much to identify the site and organizational framework where digital publishing could take place, but to identify the work and its relevance, given the nature of the material and the role played by communication vehicles for research results in the scientific community.

The publishing process, in a broad sense, covers manuscript selection, processing, formatting, dissemination and preservation. In the journal sector, the first of these functions is largely assumed by journal editors who control the editorial process. The other functions are more part of the publishing and dissemination processes and are often assumed by professionals, although journal editors who are major players in those areas do exist. Digital publishing, as we have pointed out, does not marginalize publishing work as a whole, wherever it is practised (professional publisher, independent journal, library or other). Instead, it enriches it and poses challenges that can be met by turning to new resources as opposed to the resources involved in well-known print processes. Technical aspects and equipment are less at issue here than are human resources with the training and expertise to control particular publishing processes and to produce, learn and develop applications concurrently. That is why it is important to count on a core of qualified individuals who participate fully in the development of practices, procedures and conventions.

The value added by the publisher, in the generic sense of the word, in digital publishing assumes an investment in infrastructure design and development, including the design, expertise and procedures needed for high-quality, effective digital production and dissemination. For a small independent publisher, a journal and even a publisher with a limited volume of activity, resource needs pose problems. Gathering the fairly diverse resources can thus be, although possible, very challenging, in particular for single journal producers. Without prejudging the nature of the organization recommended by promoters, it is important to consider the infrastructure components to be implemented and the functions to be performed and to find ways to share resources and expertise. These concerns need to be addressed by journal editors intending to become involved in digital publishing as well as by promoters of collective journal sites.

SERVICES

The services that play a part in a digital journal publishing project include intellectual, organizational and material infrastructure components.

The *publishing* service based on the integrated XML model or the page layout to XML model, in its most ambitious form, accomplishes all of the following tasks: receiving manuscripts, preparing copy, preparing proofs, digitizing illustrations and photographs, entering corrections and preparing the digital version and page layout for the print version. In the process, metadata are produced, ideally automatically, from the processing chain.

The value of a collection of journal articles lies in the quality of the articles, but also in the number and accessibility of journals and articles on a given subject. That justifies *retrospective digitizing* of previous issues, a process that involves producing digital versions of the desired articles and creating metadata to allow readers to search current and retrospective collections of a journal using the same search engine.

The dissemination system consists of an access interface with the objective of making a collection of articles available by subject first. The proposed components for implementing *dissemination services* are the creation of a dissemination platform, the use of a metadata model, a tool for identifying and permanently referencing each of the articles, and, potentially, a service that allows bibliographical references to be linked to the articles cited (e.g., *CrossRef*). Additional means are provided by selective dissemination of information (such as tables of contents of journal issues and abstracts) to interested individuals and subscribers to the publication, taking into account the performance of various pages in the most frequently consulted search engines and the possibility of disseminating articles through various bibliographical databases that allow full-text links.

Because a model and practices have not yet been established on a national level by national libraries, journals have to adopt an *archiving service* that meets the standards set by librarians and archivists. Whatever shared responsibility there is for archiving, journals have to ensure the preservation of the digital version to guarantee access throughout the cycle of use.

Access to journals in digital form may be free of charge or for a fee. In the case where a fee is charged, a *subscription management service* has to be established.

INTELLECTUAL RESOURCES

First-rate intellectual resources are needed to perform the various functions. For a single journal, the resources may be reduced to a single person, who is particularly able and multi-skilled and whose commitment is unfailing. For several journals or a group site, specialization in tasks is possible, if not desirable, so that the team combines complementary qualifications. Whatever model is chosen, the interaction of various professional and technical competencies allows services to be designed, implemented and delivered under appropriate conditions, while accommodating changes in standards and practices as well as taking the available financial resources into account. That requires an intellectual vision aligned with the production conditions and professional management in step with a service delivery program.

Four aspects can be highlighted:

1. Greater visibility, an increase in the number of users and the impact of the site on the international scientific community depend on the capacity to create an intellectually dynamic site that sustains, welcomes and stimulates activities able to attract the national and international research community. The best promotion for a site is to make it a *dynamic intellectual centre* that provides space for activities and offers first-rate services. That requires substantial resources in terms of both quality and size.

2. Digital publishing is dominated by innovation. The technology used and Web customs and practices evolve rapidly and require continual questioning of expectations and forecasts for digital publishing and dissemination. Production procedures, user interaction, software and other tools are thus far from stable in the digital publishing environment. *Technology watch* is fundamental, given the evolution in practices and standards for format, production and dissemination tools and the identification and transformation of both technical and economic dissemination models.

3. The need to be at the forefront of technology makes it necessary to regard resources for application research and development as the lifeblood of the project. We cannot escape the need to invest in applied research, experimentation and improvement of dissemination functions, particularly if we want to participate, in the near future, in stabilizing the production process and developing services that reveal the full value of digital publication by exploiting its functionalities. The expression "small r big D" reflects that mission, because *development and integration of applications* are essential for production and dissemination.

4. The great mobility of human resources in this sector makes it necessary to identify conditions that ensure redundancy in expertise for managing, administering and maintaining information systems. In this context, *recording and sharing information on procedures and decisions* become highly important, a practice that ensures the stability of the information systems developed and rich documentation for team members.

TECHNICAL AND HARDWARE ENVIRONMENT

Digital publishing of journals is only possible if we combine the conditions for secure, stable and ongoing hosting services for the production, dissemination and archiving of documents, features that are both technical and organizational.

A site that hosts one or more journals contains sensitive and strategic data and should be installed on its own server. The needs and environment of the host site, with its hardware and software components, are defined based on identifying a series of factors, such as:

- secured hosting and server administration services;

- the estimated disk space required for journal article content (including current publications and retro-spective digitization) and for the operating system and various dissemination software packages;

- the capacity to add various functionalities to the dissemination system, such as a full-text search engine using metadata, personalized access for visitors based on their interests and entitlement to access, recording user habits and adaptation of the display to the reader's hardware and software preferences profile.

Hosting should include, in particular, access by the server to a very broad bandwidth and a strategic position on Internet interconnectivity networks, constant monitoring of equipment and continuity of service. It should also include an air conditioning and cooling system, guaranteed power supply, backup mechanisms and DNS maintenance to determine the logical names of servers.

The technical environment complements the human resources to be mobilized and is a necessary condition for implementing a production, dissemination and archiving structure for digital journals.

A CONTROLLED TRANSITION PROCESS

There is every advantage in approaching the transition pragmatically, which means maintaining a constant tension between the change process and the constraints that shape that change. The tension is dynamic, polymorphic and follows an irregular rate, and a key quality is sustainability. One of the issues identified at the outset, i.e., parallel print and digital versions, has been dealt with from that angle.

Material produced using the four processing chains described in this chapter, accompanied by metadata, meets the conditions necessary for online dissemination, although with different use levels and longevity indicators. The decision to use one model or another is based on operating conditions, financial means and available resources. More than one option is justified, depending on the objective for adopting a digital version of a journal, from straightforward online dissemination to a complete transition to digital publishing.

The experience acquired and services we provide or have access to already allow us to identify operating methods that maximize the resources to be invested in such projects. The essential objective for journals is to promote the implementation of a new processing chain that is able to both control the transition and to facilitate adopting the digital version as the primary version by making choices that ensure information management and longevity. To meet that objective, we have to distinguish what is essential, i.e., what is indispensable to achieve the path chosen, from what is secondary, which often takes the form of methods, procedures and stewardship issues. Although far from creating any certainty about the path mapped out, our acquired experience enables us to affirm the need for maintaining sustained tension toward clear objectives for the shift to digital publishing, while practising the cardinal virtue of flexibility and promoting activities that contribute to network development.

5: CONVERSION to DIGITAL TECHNOLOGY: ACTION by JOURNALS

Adopting a digital version is not generally seen by a journal editor as replacing the print version. Most of the time, the electronic publication is regarded as an extension of the regular activities of the journal. The transition and transformation at work are supported by institutions (along with the communities that run them) that are thereby demonstrating a capacity for initiative and openness to change, all the while remaining rooted in the proven operating methods still in use. This is a social process which, although driven forward by the renewal of scientific communication forms and the initiative of its players, drags along the weight of recognized, legitimized practices, socio-institutional burdens and even behavioural inertia.

The capacity to take effective action to transform the conditions for the survival of scientific communication journals in the digital era is less an issue of bypassing journals than considering the conditions for their survival in order to maximize the impact of a transition and renewal strategy. The more we acknowledge and positively anticipate development toward digital publication and dissemination, the more we need to examine the reasons that motivate it and the methods for adopting that option. In that sense, an understanding of the social process at work is needed to target and synchronize the method of intervention. It is just as necessary to plan the transition and organization methods that can counteract current fragmentation and establish major centres for the convergence and dissemination of knowledge.

In this chapter, we deal with the conditions for the shift to digital publication and dissemination, highlighting the underlying social dimensions. The process can neither be reduced to its technical dimension nor, from an entirely different angle, be represented as a manifestation of intentional action. That would minimize the key condition for success, which is action that forces and stimulates change in the real situation, with an understanding of the key factors that act as constraints to that action. To open the discussion, considering the benefits associated with digital publishing allows us to highlight the social dimension inherent in any transition.

PUBLICATION DELAYS AND THE INTRODUCTION OF DIGITAL PUBLISHING

When we deal with the renewal of publishing conditions through the introduction of digital publishing, one of the issues raised is publication delays. Publishing and disseminating an article in the digital world should *a priori* take significantly less time than is currently allotted for the process from a submitted manuscript to a printed, distributed article. However, when considering the process leading to the publication of a digital article and the estimated time needed, it becomes clear that a purely technical point of view obscures a significant number of the components of the process, components that are predominantly social.

In scholarly book publishing, the publisher is the only one, or the first one, accountable for publication delays. For journals, the editorial work is largely done by the journal's editorial team, so the publisher can only be accountable for a limited portion of the time from the date of submission of a manuscript to the date of publication. In the case of Geoscience and Remote Sensing Society publications, the average period of time from initial submission of the manuscript to its publication is estimated[1] at 21.8 months. Yet less than five of those months are spent on programming, composing and producing the journal. That means the evaluation and review process, i.e., editorial work, consumes a major part of the time in the chain leading to publication. That proportion is, in fact, entirely appropriate for journals in the humanities and social sciences.

In addition, researchers hold paradoxical values, at least in their outcome. A survey of the quality criteria for journals conducted by the American Geophysical Union shows that readers appreciate the scientific content of a journal most and pay careful attention to the quality and correctness of the writing, criteria that require a process of selection, review and correction that takes time. Concurrently, they are concerned, to a lesser extent but significantly, with the speed of publication. Such diverse priorities are difficult to reconcile optimally.[2] The time issue is becoming increasingly sensitive, which explains the growing interest in preprint servers. Researchers in the humanities and social sciences are no strangers to these types of concerns. They may not see publication delays as a direct attack on the relevance of their message (although that may be the case), as much as they feel frustrated and see delays as interfering with their legitimate desire to see the article disseminated as soon as possible.

The introduction of digital publishing is likely to reduce the delays. But how much? As long as manuscript evaluation, selection and correction procedures are not substantially modified, recourse to digital communication will not dramatically shorten publication time. It is no coincidence that the established peer-review process is the topic of lively discussion in favour of the potential offered by digital publishing.[3] Whatever the outcome of the debate, that practice is sure to evolve as institutional conventions and recognition are renewed. Digital publishing is seen here more as a potential condition than as a necessary and sufficient condition for such change. The time saved by the introduction of new procedures still remains to be assessed.

In the current state of affairs or some other context, we can envisage a digital system that can monitor the complete editorial process.[4] The introduction of digital publishing activities should include reorganizing the various stages that make up the processing and operating methods from submission of the articles to their dissemination in a digital version. Reorganization requires learning that is appropriately shared among journals and may lead to group action, such as the acquisition of a software system for monitoring the whole editorial process for the use of several journals. The effects of the introduction of digital processes will certainly not be instantaneous, and the rate of adoption is hard to predict for the moment.[5]

Without necessarily going quite that far, numerous journals regularly use digital communication to simplify and improve the effectiveness of editorial procedures. That use is destined to grow and affect print journals as much as digital journals. While digital communication seems to be applied mostly to the editorial work in journal publishing, it is also clearly a mode of communication commonly used throughout the academic community (therefore essential in the practices of any journal editor).

In any event, just because we gain from the "speed of machines" does not necessarily mean the "slowness and vulnerability of human faculties" adopt a pace to match what is technically possible.[6] That slowness and vulnerability are reflected in a whole series of manifestations from stretching activities out over time to the time it takes for reflection and writing a review. When we give a reviewer a month to provide an opinion on a manuscript, we know that a few hours are enough; however, we do not have any control over the expert's agenda or when he or she will choose to do the job. That observation illustrates the real time frames that are hard to compress despite technology, and we should not assume the necessary progress in practices and organizations from the technical progress.

In digital publishing — as distinguished from electronic communication — we can easily imagine that the introduction of digital methods speeds up operations in the production of articles, allowing articles to be published and distributed as soon as the final version is prepared and reducing constraints associated with the scope, length and iconographic complexity of articles. All those elements could have an impact on reducing publication delays; however, we still need a fairly clear idea of the role played by the publisher and a redefinition of the operating methods in digital publication.

THE QUEST FOR LEGITIMACY

On another note entirely, institutional recognition of digital publications is often perceived as a significant issue, because lack of consideration of electronic publications in career review and promotion procedures undermines the value professors see in them. The relative novelty of digital publications may spontaneously elicit the distrust of peers. Should we therefore assume there is a lack of consideration that adversely affects researchers?

We already know that, in some sectors where the scientific network is not extensively or well structured, preprints play a central role in knowledge dissemination and enjoy institutional legitimacy.[7] Those sectors have been particularly favourable breeding grounds for the emergence of digital structures for communication of research, and their contribution is unquestioned.

On a broader scale, the issue of institutional recognition of digital publications does not appear fundamentally different from the issue of the criteria for the qualitative assessment of conventional publications. However, in the absence of direct discrimination, we might ask whether some factors have systemic effects that work against that type of publication.

Institutional review procedures for promotions only partially allow an assessment of the intrinsic quality of publications. In North America, at least, the procedure generally involves soliciting review reports from specialists in the field. The overall assessment is still largely determined on the basis of indicators of the recognized quality of the selections made by journals, publishers and so on, their rank in the hierarchy of the scientific communication system, impact, distribution and other indicators.[8] A report on this issue, written in 1999 for New Jersey State University,[9] points out that the overriding principle in reviewing publications for promotion purposes remains quality control, i.e., the rigour of the selection criteria, peer review and editorial refinement. In general, we conclude that what counts is the quality and assurance of wide distribution of the contribution, in the present but also in the future, whether the medium is print or digital.

Journals are institutions in the scientific community that have defined their place and stood the test of time. For many, that is the dimension considered in professional reviews, so an exclusively electronic publication, which by definition is recent, has to fight, at least for a while, certain "structural" deficiencies, as is the case for any completely new journal.[10] Among the deficiencies are the difficulties encountered in demonstrating scientific rigour, in assessing the identifiable impact of the publication, and in indexing the journal in major authoritative indexes. New digital journals are also often the target of scepticism among established researchers who are reticent to publish in this medium, particularly in the humanities and social sciences. Yet, at the same time, the new media have the greatest propensity to attract younger, less experienced contributors.

All those factors may cast a shadow on "digital vehicles" in reviews.[11] On one hand, those factors may be considered a barrier to the entry of new journals, especially because the scientific communication system is doubly structured by the scientific community and its hierarchical networks and the economic and commercial organization that defines the publication and distribution framework.[12]

On the other hand, those factors may be seen as constituting a selection mechanism that is intended to be neutral yet effectively introduces a prism (which is also debatable) whether the publication medium is print or digital.[13] At the same time, we cannot ignore that to some, a journal distributed free of charge on the Web, and therefore with no subscribers, appears less serious than a journal with a fee and a searchable index for consultation, proven to be of "serious" interest to institutions and potentially to individuals. It is as if financial expenditure could bear witness to how well-known a journal is; naturally, that is a matter of perception, selected indicators and subjectivity. It does not mean journals distributed free of charge cannot achieve an enviable standing, quite the contrary.[14] However, their standing would originate within the discipline and be corroborated by other indicators.

The shadows thus cast on new digital journals clearly dissipate when an already well-established journal of good standing and reputation is published in parallel print and digital versions, or even in a digital version only. The barriers mentioned that impede the entry of new journals normally have little or no impact on journals that have already made their presence felt in the research community and their conventional print versions have proven influential.

We may question the need to produce new digital-only journals for the same niche, if print journals that are not prey to oligopolistic commercial groups and have played a significant role in their field until then undertake the process of transforming to digital publishing. We can expect the print version to be distributed in parallel, for an indeterminate period, although the digital version may gradually become the first reference version.[15] The routine presence of digital publications in researchers' practices should change perceptions, so that the issue of institutional recognition will no longer be influenced by the distribution media.[16]

Of course, the issue of quality criteria and their use in promotional procedures can be discussed. The introduction of new information technology, more widespread use of the Internet and the practices introduced by both may stimulate a critical review of those criteria. Rethinking the standards in effect is always desirable; however, it should not lead to the assumption that contesting the criteria is associated with the inability of digital publications – with or without a print version – to meet them. Although the publication and distribution media impose some changes in the indicators used, they remain fundamentally neutral in terms of quality, scientific rigour, dissemination and longevity.

THE ADOPTION PROCESS

Let us start with a straightforward observation. The implementation of digital publication and the disappearance of the print version of journals cannot be dictated any more than changes in social practices and disruption in institutions. That observation is not original in itself; however, unlike instantaneous utopias, it assumes that we have taken the trouble to understand the scope and limitations of the scholarly publishing environment to ensure the relevance of the action planned while optimizing its impact.

FRUSTRATED EXPECTATIONS

A number of examples show that the route taken and the time it takes to complete projects — requiring change and renewal in practices, organizational methods and communication networks — often lead to shifts and detours that frustrate impetuous undertakings. It is not a question of inertia or passivity, simply realism and awareness of the conditions for their completion. A few examples illustrate that idea and, most important, engage us in continued reflection on the social dimensions of the transition to digital publishing.

Recounting the experience of the *Journal of Biological Chemistry* (JBC), Tom Abate illustrates how long it takes to change practices. The editors produced a digital version of the journal and proceeded to encourage the library at Stanford University to establish a non-profit digital publisher — HighWire Press — for the purpose of publishing the journal online. (The journal was then published in January 1995, followed by many other digital journal publications.) At first, the JBC editors thought the digital version of the journal would rapidly supplant the print version, making it superfluous. Contrary to all expectations, however, the demand for the print version has remained particularly strong. That slowed the process anticipated by the journal's editors. Noting that phenomenon, the editors said they were convinced the transformation would occur, without being able to set a deadline, despite their resolute position. Robert Simoni, the key promoter of the transformation, acknowledged that those central to establishing the digital journal overestimated the speed of the transition.[17] What should have taken a few months took the journal a few years, with particularly resolute and engaged editors, and the process is not over. HighWire is one of the major players in the field today.

Rosenblatt and Whisler,[18] respectively a librarian and editor at the University of California, pointed out in 1997 that the pressure to maintain both the print and digital versions of journals at the same time will be intense in the near future. There is evident reticence on the part of libraries and their users to abandon the familiar archiving format, i.e., the print version; moreover, investments in the technical infrastructure required to make digital documentation available have been lagging behind. For publishers, converting to new processes and uncertainties about marketing digital documentation and readership demand argue in favour of a transition period during which the print and digital formats are produced in parallel. Today, with a few years distance, we see that the shift to digital publishing has taken root while dealing with those points of friction.

Our experience with *Érudit* and the studies we have done of the process – based on interaction with journal editors and the academic community – confirm those assessments. A whole series of variables have slowed the emergence of new practices that would have decidedly promoted digital publishing. Everything that is happening in the transition period suggests that some players want to maintain the best of both worlds by hesitating to cross the Rubicon.

That period cannot go on indefinitely. Rapid change in the environment that affects scholarly documentation (including writing, production, distribution, access and use) renews the issue of maintaining the print version. Yet, does this oblige us to turn the page without further adjustment? Motivated by the best intentions and driven by a sense of history to adopt digital publishing, ill-informed action may produce results other than those sought. In winter 2000, the Quebec public research funding agency – the Fonds FCAR – decided that assistance granted to journals would be calculated (after a maximum three-year period) exclusively based on the cost of the digital version. That measure had the advantage of clearly setting an objective and a path, but was received by the journal community as an intentional action that did not take real publishing conditions into account. The policy was then revised to support the adoption of digital versions that would not result in the disappearance of the print versions, a policy that is undoubtedly more appropriate for producing the desired results.

Judging from developments over the past few years, we estimate that simultaneous print and digital versions of journals will be necessary for an as yet undetermined period. At the same time, we have to recognize that time is passing, noteworthy transformations are underway and the transition period is not nearing an end. That is the shifting reality we have to grasp.

WHAT DO JOURNAL EDITORS THINK?

In a study of the scope and limitations of establishing a digital journal production and distribution site in fall 2000, we held meetings to consult with the journal editors of half the publicly funded journals in Quebec. Significant trends were revealed in the course of those discussions that are worth considering, especially because they diverge from the simplistic "resistant to change" label often applied to journal editors. In addition, various considerations are fairly universal in the humanities and social sciences non-profit journal community involved in national infrastructures for knowledge dissemination.

Editors have a well-honed knowledge of the conditions for survival of their journal, from editorial work to subscription sales, right through technical production and the journal's financial situation. Journals are edited by professors who are generally very active professionally and have experience as national and international authors. They show a definite interest in the evolution of scientific communication methods and understand the key issues.

Digital publishing does not aggravate journal editors, quite the contrary. Information and excerpts from most of the journals are already available on a Web site and, in some cases, a digital version of the journal already exists. The editors recognize that digital publishing is the way of the future and the transition is essential to maintain and increase journal visibility. From that perspective, some considerations appear essential for ensuring the transition to digital publishing with the co-operation of journal editors.

To get things done, realism is essential. No doubt it is inappropriate for the shift to be imposed by intentional action for immediate application. As fascinating as the idea may be of considering the digital version the primary version of the journal, it involves a change of culture, content review and modification, and a redefinition of working methods. To meet those demands, a multi-faceted transformation strategy has developed, a strategy that affects authors, editorial and administrative teams (existing and restructured), the financial capacity of each journal and a new understanding of the journal's technical production and digital dissemination activities. A complex transition at an accelerated rate seems brutal, at least to many, and could have adverse effects due to possible negative reactions. That is why we need to support the idea of a controlled transition.

The major challenge is to reconcile independent publishing and establishing a site for the delivery of journal distribution services. The interface can be achieved if the roles are well defined, ensuring that the site does not encroach on the editorial process and the overall responsibility for the journal, including financial responsibility, remains in the hands of the current editors. It has to be clearly recognized that the journal world is a community of established specialists and researchers, but also a community of artisans and volunteers who have mobilized great resources in the past to improve journals and increase their dissemination. The commitment that community shows for journal production and distribution attests to the individual investment and social fabric behind those types of organizations. Disrupting the commitment of the community would elicit predictable and legitimate social reactions. Recognizing that reality, a non-profit organization offering technical production and digital dissemination services is seen by journals as an attractive option, precisely because a quality digital version can hardly be produced by artisanal methods as is the case for the print version.

It is not out of nostalgia or mere concern for their own image that journal editors underline the still major position of the print version. Substituting the digital version for the print version is not a move undertaken by the major journals in each discipline. For established journals, the adoption of a digital version has not generally sounded the death-knell of the print version. The reasons cited mainly relate to the gradual change in readers' habits, concern for increasing distribution where the print version remains a highly relevant medium and the expectations of readers outside the academic community who do not have the same digital communication infrastructures. Moreover, the print version is not just a source of cost but also of revenue that exceeds the cost. Such considerations do not militate against the digital version but point out the value of maintaining a print version of journals.

Journals are vividly experiencing the financial instability of the current situation. With reduced and underfunded teams, their ability to maintain high-quality content and recognition in the scientific community both locally and abroad is currently compromised. In the short term, savings arising from the digital production and distribution of journals are certainly not taken for granted. The ability to earn independent revenue (through subscription sales and single-copy sales) and to diversify grants and other funding is now seen as a guarantee against too great a vulnerability to changes in policy by a single public institution. The idea of free access through a public sponsor is of interest in itself, but increases dependence on public funds and practically completely eliminates the existing flexibility to deal with the more or less enlightened decisions of public organizations over which journals have little influence. The involvement of a public sponsor should be subject to a financial pact with the journals to secure stability for journals and serious guarantees of public sponsorship over the long term.

Journals are institutions that have actively participated in the development of the sciences in their society and in national and international communication of knowledge. Journals have an intellectual history and are valued vehicles for the development and influence of the work of national researchers and researchers from other countries. In that sense, journals play a role that is not specific to any particular society. Journal editors feel great responsibility for protecting and developing our collective heritage. They therefore seek to avoid uncertain decisions that could turn into major intellectual and organizational costs and result in irreversible situations. In that frame of mind, they are little inclined to a spirit of adventure, not with respect to digital publishing, but more with respect to a decision-making process in which all the factors have not been taken into consideration.

INERTIA OR COMPLEX SOCIAL DYNAMICS?

Caution toward, and even opposition to, the disappearance of print media in journal publishing merits attention. The force of inertia we think we detect in those attitudes and frequently associate pejoratively with resistance is primarily an indication or manifestation of a process of professional and social adoption of innovation in scientific communication. The shift from the print to the digital version of journals evidently involves changes that are not just technical.

The scientific communication system, before being a technical infrastructure, is an institutionalized social system, with its conventions, rites and certainties. Changing the behaviour and expectations of various players and institutions is a process that meets resistance and inertia. Yet that resistance and inertia are not *a priori* the work of negative or backward-thinking minds, although there are some! We may well have a vision and specific objectives to carve out a direction and stimulate change in the forms and media journals use for scientific communication. However, it is also important to establish a consistent strategy that takes the complex process of adopting innovation into account.

That issue, which arises from a distinctly pragmatic observation, requires in-depth examination from a sociological perspective. Without pursuing an examination of that type here, it seems worthwhile to recall some thoughtful comments from the book *Où vont les autoroutes de l'information?*, edited by Marc Guillaume.[19] According to the authors of that collective work, if we are tempted to see refusal or slowness to adopt an innovation as a sign of resistance on the user's part, it is undoubtedly because we are haunted by the utopia of the glorious future of information technology. Moreover, the notion of resistance is merely seen to reflect the idea of opposing a movement or force and, in that sense, appears to be purely descriptive of an attitude. Too often, however, what we still implicitly mean is that it is the attitude of a troublemaker who opposes development we would like to be massive, rapid and unavoidable. Technology does not exist apart from its history, nor an individual apart from a social situation; on the contrary, there is strong mediation between them.

Mediation takes place through "*players* who act as an interface between designers and users — precursors, forerunners and opinion leaders are three of the best-known but certainly not the only figures. Further mediation is brought about by *institutions* where ownership is defined by the rule of law, social statutes and power relationships. Finally, there is mediation through *culture*, in other words, through social image, imagination, language and shared frames of reference."[20] All those spheres of mediation define specific contexts for ownership in which social groups are engaged. That means the process of social ownership or adoption of innovation, as a complex dynamic, directly raises the issue of identity, values, social direction and customary standards.

The transformation proceeds less through *tabula rasa* action and much more through incorporation, by sedimentation, of new practices, images and organizations into a whole that shows, at least in part, some stability in reference principles. Journals are institutions in scientific communication networks that have established their personality, seriousness and reputation. While we recognize the value of creating new digital-only journals, it also remains entirely appropriate to work with existing journals. Taking their current practices and conventions into consideration, change can be instigated by phasing the introduction of digital publishing.

Renewal in the design, production and distribution of the journal can be achieved without radically questioning all the associated editorial and physical benchmarks. That is why the print and digital versions are most often published simultaneously, why the concept of issues and volumes is most often maintained, and why, where thematic issues are produced, that practice is retained and so on. That should not be taken as an argument for the unchangeable nature of journals, quite the contrary. We should not confuse the cumulative dynamic of change with stasis; we are carried by that dynamic toward something else, at a pace determined by the conditions for achievement and results.

In an article on the legitimacy of digital publications (based on the experience of the *Journal of Artificial Intelligence Research*), Kling and Covi[21] portray scientific communication as a sociotechnical system. The reception and legitimacy of an digital journal, they argue, are certainly conditional on the use of digital functionalities beyond distribution, but also largely on the retention of print media in parallel. In addition, they emphasize the presentation of a digital version that uses all the conventions of the print journal. Should the concern for an exact replica become a cardinal virtue in the transition underway? That may seem dubious (as it certainly does to us). Without trying to vindicate the exact replica, we cannot ignore the fact that the existence of a print version is one of the conditions for the success of digital journals.[22] Without a doubt, that illustrates the constraints change strategies have had to face to ensure a smoother transition and to achieve a certain level of effectiveness.

ORGANIZATIONAL CONDITIONS FOR CHANGE

The transition to digital publishing takes place in the socio-economic and scientific structure in which journals and scientific teams are organized. The relative position of the structure in scientific communication networks determines resources and authorizes prospects for development. As we move away from national structures with little or no involvement in the top ranks of knowledge dissemination networks of journals, the resources and range of possibilities both shrink. Considering that situation, which was described previously as well, raises the dilemma of the technical and organizational direction of the shift toward digital publishing and distribution of journals, in particular in the humanities and social sciences.

ELECTRONIC PUBLISHING AND
ORGANIZATIONAL FRAGMENTATION

Due to the organizational and institutional dispersion of journals, the digital publishing challenge may lead to reinforcing and pushing the fragmentation in journal operating methods even further and result in developing digital solutions on an individual basis within the limited resources each journal has at its disposal. Technically, some solutions appear sufficiently easy to manage and may lead to making a digital version of the journal available on the Web, free of charge most of the time and offering rudimentary services. With those solutions, however, it may be difficult to achieve publication, preservation and distribution appropriate to the journal's mission and the nature of the articles published.

In those cases, a whole series of questions remain and fairly quickly reappear once the euphoria of the launch has faded, for example:

- Can the service be maintained on a regular basis for several years?

- How can regular file updates be ensured with a proprietary format chosen for cost-saving reasons?

- Is it possible, when operating individually, to cover the cost of using and developing value added services, such as cross-references and selective dissemination of information?

- Can we guarantee readers ongoing quality access, which means rapid response time?

- Can we ensure the security of the server and the network environment for distribution service?

- Can we establish an effective distribution system, with quality services and adequate visibility, despite the overabundance of sites on the Web?

- What economic model has been selected? Can it be implemented if it is based on filtered access by subscription? What methods are available for developing institutional and individual subscriptions to digital media?

- Can we collect revenue from consultations?

- Can we handle the responsibilities and tasks involved in the long-term preservation of the digital version of journals?

- Can we maintain the technology monitoring and development needed for digital publishing?

- Given the choice of digital format, are we actually progressing toward the creation of a genuinely digital journal or just transposing a journal essentially designed for print?

Those are some of the many questions that show it is technically possible to manage a journal digitally without necessarily being able to proceed with the publishing, preservation and distribution appropriate for scholarly journals and, consequently, for the nature and mission of the material.

SEIZING THE OPPORTUNITY

The challenge of digital publication can be viewed as a major opportunity, not only to improve the conditions for dissemination but also to rethink journals and methods of formalizing and communicating research results. The challenge is to adopt information technology to serve scientific publication. Of course, new practices will completely change the conditions for research and the distribution of its results. In that context, it is important to show genuine concern for aligning the value of the published and distributed content with good publishing practices and technical choices. The impending transition also affects the upgrading and use of research results, project communication and dissemination, research tools and conditions, the publisher's role and the components of the scientific communication chain − from authors through publishers and libraries to researchers and readers.

The challenge encourages questioning the organizational structure of a publishing practice that is characterized by dispersion and fragmentation in light of the now prevalent tendency toward networks. Can the isolation and individualization in journal publishing practices be considered favourable to active participation in the changes underway in the scientific communication system? There is no simple answer to that question; however, we may see the challenge as an opportunity to rethink how publishing and distribution tasks are structured.

It is desirable for publication and dissemination activities to be a bigger part of the vision of change in scientific communication and considered with well-honed knowledge of the functionalities of information technology applied to the journal mission. Publishing and distribution activities contribute to journals in the more technical phases of production and by making journal articles available. Editorial autonomy, the peer-review process and the selection and correction of manuscripts are the prerogative of journal editors. Defining the purpose of a journal publication or dissemination platform makes no inroads into the editorial autonomy, administrative independence and intellectual and physical personality a journal has acquired. Journals are being asked to delegate direct production and distribution tasks for the digital version to a service provider who must work closely with the journals to upgrade content and increase impact by systematically exploiting the Web's potential.

A service delivery site is the first constellation in a network that can only develop further. The network concept is key in describing the strong trend in the transformation of social communication practices, in particular.[23] The word "network" has various connotations, depending on the context in which it is used, for example, a societal project, an unavoidable constraint or a relentless reminder of social organization. We use it in a strategic sense.

Can the isolation and individualization in journal publishing practices be seen to lend themselves to active participation in the changes underway in the scientific communication system?

Certainly, it is possible to envision the challenge of digital publishing pushing us toward further fragmentation through the adoption of partial solutions. However, we may also see the challenge as an opportunity to rethink the fragmented structure of publication and to opt for formulas that promote inclusion in networks for publishing and distribution activities.

By taking all the factors that affect journals into account (particularly in Quebec, Canada and France), by referring to major distribution sites (particularly in the United States) and by considering the issues previously identified, we are in a position to roughly draft an outline for a dissemination platform for journals in the humanities and social sciences. The plan would chart the main path, select specific strategic options and proceed with technical choices.

THE NETWORK ISSUE

The network issue is distinguished by a few important processes that have to be carefully examined. Manuel Castells[24] observed that "the dominant functions and processes in the information era are increasingly organized in networks. Networks are the new social morphology of our societies, and the widespread networking logic largely determines the processes of production, experience, power and culture.... What is new today is that the new information technology paradigm provides the physical basis for its extension to the whole social structure." Among the components of the paradigm, we note the omnipresence of the effects of new technology, the network logic in any group of relationships that uses new information technology and the increasing convergence of particular technology in a highly integrated system.[25] By observing the strong trends in the evolution of today's societies that are influenced by information technology, we can identify the orientation and even strategies for action.

STRATEGIES FOR ACTION IN PHASE WITH THE ENVIRONMENT

It is difficult for isolated journal editors to find the means to ensure the digital publication and distribution of three or four issues a year at reasonable cost under reasonable production conditions. The only possible way to manage that would be with a multi-skilled volunteer – an alternative that can hardly be seen as a stable and generalized model over the long term. At first glance, gathering the necessary resources and ensuring visibility and effective distribution appears to be conditional on a high-volume operation. We may raise the question of a breakeven point in conventional terms. However, the "technical" variable should always be interpreted in the context of the organizational framework that would allow quality publishing and a strong presence in international networks.

A host structure modelled on the principles of a non-profit corporation appears more conducive to any concerted effort and appropriate for a service delivery site. It is certainly the form most suited to the scholarly mission. With an effective organization, we may even see it as the ideal springboard to counterbalance and even compete with commercial oligopolies that publish scholarly documentation. In any case, it is the environment that can best provide quality service at acceptable cost.

The major and relentlessly unavoidable challenge is to implement a process that is rooted in the journal environment and takes its characteristics into account, while recognizing, respecting and maintaining the autonomy of journals from an editorial, financial and organizational standpoint. To that end, we have to recognize that the editorial process of journals is largely independent of the publishing work as such. Moreover, it is important to distinguish between the participants and the organizational sites, because journal editors can perform all the editorial tasks and take overall responsibility for the journal, while delegating publishing and distribution tasks to others.

For action to be relevant and effective, it must be based on a detailed vision that transcends the macroscopic level and broad generalizations and grasps the structure of communication channels in their economic and sociological dimensions. This is an opportunity to propose an economic model other than an oligopoly. In addition, considering the second factor in the journal publishing structure, i.e., sector-based scientific communication practices, it is possible to implement an action plan that is in step with the players already in place.

Learned societies and university presses have a significant position in journal publishing in the natural sciences and the humanities and social sciences; that applies to both the number of journals published and their importance and quality. Moreover, it holds true in the select sphere of journals that have the most impact worldwide, i.e., the top twenty-five journals in the disciplines, and even more so for journals that are part of national scientific communication infrastructures. Non-profit publishing sites are major players. The organization of publishing shows that the practices of non-profit publishers do not contribute to the financial crisis in scholarly publishing, and they play a primary role among dominant journals. Generally, the practices of small private publishers that are active in the journal sector are fairly similar to the practices of non-profit publishers. We can capitalize on that base by pulling together the resources to make journals part of the new scientific communication environment.

NETWORKED PUBLICATION SITES

From the viewpoint that networks are an organizational method throughout society, a policy of both concentration of resources and networking is of strategic interest for developing an alternative to oligopolies. We therefore have to carefully consider the possibility of developing digital and print publishing and distribution centres dedicated to the scientific community and the research community.

First, as an initial form of journal grouping, a distribution site has the primary responsibility for ensuring the impact and visibility of the collection on the Web. We have already discussed distribution strategies consistent with that mission (Chapter 3). The strategies have to be supported by a package of services to journals and readers, such as a secure host, digital dissemination of retrospective and current issues, metadata and permanent referencing, an appropriate high-performance search engine, access filtering and technology monitoring. ·

Establishing that type of site also goes hand in hand with confirming and acknowledging the editors' authority for their respective journals. Sovereignty issues thus clarified, the site can then benefit from relying on the editors to identify and develop services for both journals and users. By delegating only the responsibility for the digital production and distribution of the journal, the editors maintain intact their control over the journal's destiny. That respects the journal's dynamics and supports the work of the editorial teams, while providing them with the means to exploit the product of their activities digitally. The concerted efforts and resulting resources of that mode of organization provide journals with means well beyond the capability of each individual journal, allowing the journals to contribute to shaping the itinerary of the network being built.

Second, the network should form a fairly close collaborative group of digital journal distribution sites.[26] As a minimum, the distribution sites should establish common standards for the production of the metadata associated with the articles made available online. That allows the creation of an expanded database, access to the database to view articles — which does not exclude the presence of various access points — and the installation of a high-performance search engine that can scan a rich corpus made up of the publications of many journals.

Networking distribution sites would be based on shared metadata for articles hosted on each of the network sites. It would therefore be possible, at every site, to search an overall corpus of data and to direct users to sites where the articles cited in the search results are hosted. That allows the unified description of articles while maintaining the autonomy of each site, particularly in terms of conditions for access. The conditions for developing the network do not impose overly heavy constraints. On the whole, a network increases the value of the information holdings of journal sites tenfold, amplifies their visibility on the Web and facilitates strategic positioning in dissemination to search engines, directories and major indexes.

We also have to rely on a communication code (in the broad sense) and a common or closely compatible frame of reference for the services to be provided to journal readers and users. That does not imply uniformity, but rather a synchronization or alignment of practices. Furthermore, the alignment would benefit from synchronization with libraries to ensure compatibility with their procedures for processing and distributing scholarly journals.

In terms of language, a network of dissemination sites for Francophone journals would undoubtedly be a very valuable lever for the dissemination of French-language research, providing consistency and continuity. The virtual grouping of the major, essential scholarly journals of the Francophonie would provide an extraordinary showcase for communication vehicles for the advancement of research in that language subgroup of the international scientific communication system. It would thus be a fundamental way for French-language research and scientific discourse to become established on the Web.

That does not necessarily mean trying to enclose or isolate the subgroup, which leads us to consider a third level of understanding of the network. If we stand to gain by giving form to this subgroup of scientific communication, we have an equal interest in making this subgroup a springboard for entering the international, largely English-language, scientific communication system. To establish links with the international community, the Francophone network needs to expand and participate in exchange, in particular with major English-language sites. The effort to penetrate dissemination channels should be supported by a vigorous strategy and pursued an ongoing basis. The main body of data, as well as a unified method of structuring information, or metadata, are major assets in ensuring a presence in bibliographical and textual databases, search engines, directories and major indexes. Depending on the journals and their respective sectors, the targets need to be varied for greater relevance and impact; however, the dissemination process and methods selected are practically the same.

Among the conditions that ensure an active and attractive presence on the Web are, to summarize, information sharing and the possibility of searching a rich, diversified corpus while respecting the autonomy of partner sites, commitment to acquiring professional competencies at the leading edge of technology, delivery of more high-performance and individualized services and expanded distribution through a more effective and well-adapted network that works closely with institutions that are also dedicated to the dissemination and consultation of scholarly documentation, such as libraries, and major intermediaries in the form of search engines and databases.

◆ ◆ ◆

A fruitful discussion on the transformation of scientific communication and the role of digital journal publishing in particular should be based on an identification of the participating players and an acknowledgment of their respective positions. Beyond the players, describing their roles in the adoption of information and communication technology allows us to define the nature of their activities in publishing and disseminating research results. After identifying the players and their roles, it is easier to identify network-building strategies for scientific publishing and communication that originate from and serve the academic community.

Large commercial journal publishers have taken control of major journals in various sectors and, from an oligopolistic position, developed a marketing strategy that enables them both to impose their bundle of journals and to charge exorbitant prices. Their commercial power should not overshadow the more than significant presence of non-profit players; however, it underlines the fragmentation of organizational methods in scientific publication among those players. Networking strategies at various levels, in addition to setting non-profit players apart from oligopolistic practices, would unite the strength of the cohesiveness and plurality of personalities of the players present.

The roles played by various players are not embedded in an unchangeable organizational form. Publication sites in the form of independent journals, learned societies and research institutions that orchestrate all operations, university presses and, more recently, services associated with libraries have adopted various organizational forms. Those forms are currently being renewed by innovation and initiatives. Publishing and distribution mandates and practices are being redesigned but still occupy a central position in scientific communication. The chain from publishing to distribution is taking new forms and redefining its interfaces as well as the practices of some players, particularly libraries.

The ways organizational forms take shape and authors define and perform their roles do not follow any fixed codes. Can we simply hope to profit from the skills and expertise of each individual in the implementation of a network, a strategic option that will develop complementary and non-competitive expertise in a chain of centres of excellence comprising the diverse digital publishing and dissemination platforms?

NOTES

FOREWORD

1 We wish to thank Maureen Ranson for the translation and Cathleen Poehler for the revision.

2 The partners we have been working with for many months now are: Lynn Copeland and Rowland Lorimer (Simon Fraser University), Frits Pannekoek (University of Calgary), Karen Turko (University of Toronto), Gwendolyn Ebbett (Windsor University), Alan Burk (University of New Brunswick) and David Moorman (Social Sciences and Humanities Research Council of Canada), as well as our partners from the *Érudit* consortium, among them Claude Bonnelly (Université Laval) and Chantal Bouthat (Université du Québec à Montréal).

INTRODUCTION

1 <http://www.erudit.org/>.

2 This book draws, in particular, on Boismenu, Gérard, et al., *Le projet Érudit : Un laboratoire pour la publication et la diffusion électroniques des revues universitaires*, Montréal, Report on the pilot project conducted by Les Presses de l'Université de Montréal submitted to the Fonds FCAR, May 1999, 225 p., <http://www.erudit.org/erudit/rapport/index.html>; Boismenu, Gérard, and Guylaine Beaudry, "Publications électroniques et revues savantes : Acteurs, rôles et réseaux", *Documentaliste. Sciences de l'information*, 36 (6), 1999, pp. 292–305 and *Documentation et Bibliothèques*, ASTED, 45 (4), 1999, pp. 149–60; Guylaine Beaudry, Gérard Boismenu, et al., *Conception d'un portail de production, de diffusion et de gestion de publications électroniques: Étude de faisabilité*, Montréal, Report submitted to the Fonds FCAR, September 2000, 126 p.; Guylaine Beaudry and Gérard Boismenu, "Expertise technique et

organisationnelle," in Ghislaine Chartron and Jean-Michel Salaün (Eds.), *Expertise de ressources pour l'édition de revues numériques*, <http://revues.enssib.fr/>.

We wish to acknowledge the rewarding discussions we have had with other partners in the Groupe interuniversitaire pour l'édition numérique: Claude Bonnelly and Guy Teasdale, Université Laval; Chantal Bouthat, Université du Québec à Montréal; Benoît Bernier, Les Presses de l'Université Laval; and Angèle Tremblay, Les Presses de l'Université du Québec. We are grateful for their openness and co-operation, yet remain solely responsible for the proposals presented. In recent years, we have had many collaborators in research and experimental activities. In particular, we would like to acknowledge Martin Sévigny, Brigitte Gemme, Sébastien Tremblay and Isabelle Spina who contributed preliminary research for one chapter or another. Also, Jean-Noël Plourde, Michel Plamondon, Luc Grondin, Isabelle Spina and Reda Benjelloun are a competent team that promotes discussion on all aspects of the electronic publishing process. The contributions of all those individuals are gratefully acknowledged.

CHAPTER 1

1 Tenopir, Carol, and Donald W. King, "Designing Electronic Journals with 30 Years of Lessons from Print," *The Journal of Electronic Publishing* [on line] 4 (2), 1998, <http://www.press.umich.edu/jep/04-02/king.html>.

2 *Serials Price Increase Alerts*, University of Southern California, <http://www.lib.use.edu/~scheiber/price.htm>; *Price Comparison of STM Journals*, monthly, <http://www.harrassowitz.de/top_reports/index.html>; Case, Mary M., *ARL Promotes Competition through SPARC: The Scholarly Publishing & Academic Resources Coalition*, <http://www.arl.org/newsltr/196/sparc.html>. For France, see the summary table on the change in average cost of serials subscriptions in university libraries in Chartron, Ghislaine, and Jean-Michel Salaün, "La reconstruction de l'économie politique des publications scientifiques," *Bulletin des bibliothèques de France*, 45 (2), 2000, p. 33.

3 As is seen later, a journal published by a non-profit publisher costs many times less per character than a journal published by a commercial group in the same discipline. Similarly, we know that mergers of large commercial journal publishers lead to noticeable price increases. "University Press Group to Study Whether Books in Some Fields Are Disappearing: Non-Profit Journals Are Found to Be More Cost-effective than Commercial Ones," *The Chronicle of Higher Education*, 46, 1999; McCabe, Mark J., "The Impact of Publisher Mergers on Journal Prices: An Update," *ARL Bimonthly Report*, 207, July 2001, <http://www.arl.org/newsltr/207/jrnlprices.html>.

4 Abate, Tom, "Publishing Scientific Journals Online," *BioScience*, 47 (3), 1997, <http://www.aibs.org/bioscience/bioscience-archive/vol47/Mar97abate.html>.

5   McCabe, Mark J., "The Impact of Publisher Mergers on Journal Prices: An Update," *ARL Bimonthly Report*, 207, July 2001, <http://www.arl.org/newsltr/207/jrnlprices.html>.

6   Wyly, Brendan J., "Competition in Scholarly Publishing? What Publisher Profits Reveal." *ARL Newsletter of Research Library Issues and Actions*, No. 200, 1998, <http://www.arl.org/newsltr/200/wyly.html>.

7   Keller, Michael, "Innovation and Service in Scientific Publishing Requires More, Not Less, Competition," *Nature*, Webdebates, <http://www.nature.com/nature/debates/e-access/Articles/keller.html>.

8   Becher, Tony, "The Disciplinary Shaping of the Profession," in Burton R. Clark (Ed.), *The Academic Profession: National, Disciplinary, and Institutional Settings*, Berkeley: University of California Press, 1987, pp. 271–303.

9   <http://www.sciencemag.org/>.

10 <http://www.nature.com/nature/>.

11 Hacking, Ian, "The Disunities of the Sciences," in Galison, Peter, and David J. Stump (Eds.), *The Disunity of Science: Boundaries, Contexts, and Power*. Stanford: Stanford University Press, 1996, pp. 37–74.

12 Becher, Tony, "The Disciplinary Shaping of the Profession," in Burton R. Clark (Ed.), *The Academic Profession: National, Disciplinary, and Institutional Settings*, Berkeley: University of California Press, 1987, p. 285.

13 <http://www.arXiv.org/>.

14 Blumenstyk, Goldie, and Vincent Kiernan, "Idea of On-Line Archives of Papers Sparks Debate on Future Journals: Proposal by the NIH Director Infuriates Some and Inspires Others," *Chronicle of Higher Education*, July 9, 1999, p. A25.

15 See Index national des prépublications de mathématique en France, <http://www-mathdoc.ujf-grenoble.fr/prepub.html/>; CH Working Papers, <http://www.kcl.ac.uk/humanities/cch/chwp/submit.html>; arXiv.org E-Print Archive, <http://xxx.lanl.gov>; RePEc (Research Papers in Economics), <http://www.repec.org/>; CogPrints, <http://cogprints.soton.ac.uk/>; GrayLIT Network, <http://graylit.osti.gov/>; MathNet.preprints, <http://mathnet.preprints.org/>.

16 Van de Sompel, Herbert, and Carl Lagoze, "The Santa Fe Convention of the Open Archives Initiative," *D-Lib Magazine, 6*(2), 2000, <http://www.dlib.org/february00/vandesompel-oai/02vandesompel-oai.html>.

17 Harnad, Stevan, "Free at Last: The Future of Peer-Reviewed Journals," *D-Lib Magazine, 5*(12), 1999, <http://www.dlib.org/dlib/december99/12harnad.html>.

18 Le Crosnier, Hervé. *Avons-nous besoin des journaux électroniques?* Translated quotation from paper presented to Journées SFIC-ENSSIB, *Une nouvelle donne pour les journaux scientifiques*, Villeurbanne, November 20, 1997, <http://www.info.unicaen.fr/herve/pub97/enssib/enssib.html>.

19 Harnad, Stevan, "Free at Last: The Future of Peer-Reviewed Journals." *D-Lib Magazine, 5*(12), 1999, <http://www.dlib.org/dlib/december99/12harnad.html>.

20 Smith, Arthur P., "The Journal as an Overlay on Preprint Databases," *Learned Publishing, 13*, 2000, pp. 43–48.

21 On this issue, see Rowland's article, which identifies four fundamental functions of a scholarly journal: Dissemination of information, quality control, the canonical archive and recognition of authors, <http://www.ariadne.ac.uk/issue7/fytton/intro.html>.

22 Rowland, Fytton, "Print Journals: Fit for the Future?" *Ariadne*, No. 7, 1997, <http://www.ariadne.ac.uk/issue7/fytton/>.

23 de Solla Price, Derek John, "Communication in Science: The Ends – Philosophy and Forecast," in Anthony De Reuck and Julie Knight (Eds.), *Communication in Science: Documentation and Automation*, London: J. & A. Churchill, 1967, p. 201.

24 Arms, William Y. "Scholarly Communication, Digital Libraries, and D-Lib Magazine," *D-Lib, 5*(4), 1999, <http://www.dlib.org/dlib/april99/04editorial.html>.

25 Institute for Scientific Information, *Science Citation Index, Journal Citation Reports*, 1997; and *Social Sciences Citation Index, Journal Citation Reports*, 1997.

26 Becher, Tony, "The Disciplinary Shaping of the Profession," in Burton R. Clark (Ed.), *The Academic Profession: National, Disciplinary, and Institutional Settings*, Berkeley: University of California Press, 1987, pp. 271–303.

27 The impact factors for social science journals are not nearly as high as the impact factors for the natural science sectors; they also show less disparity. For all natural science-related disciplines (physics, chemistry, materials science and electrical engineering), the median impact factors are always higher than the minimum of 1.6, for physics, and the maximum is 3.0, for chemistry. The median for all the disciplines is 2.0, while the average is 3.2. The situation is very different in the social sciences (sociology, political science, social work and education), where the highest median impact factor is 1.3 (sociology), the minimum is 0.46 (social work), and the overall median is 0.97. The overall average is 1.1. The standard deviation is 0.49, which is significantly lower than for the natural sciences group (2.41).

28 This reading of reality is often seen as obvious, inevitable and unavoidable, even by those who deplore the consequences. For example, Whisler, Sandra, and Susan F. Rosenblatt, *The Library and the University Press: Two Views of the Costs and Problems of the Current System of Scholarly Publishing*. Paper presented at the *Scholarly Communication Technology* conference, Emory University, April 1997, <http://www.arl.org/scomm/scat/rosenblatt.html>.

29 The percentages are comparable to the percentages reported by Abate; 40% of scientific journals are produced by commercial publishers, while university presses and government agencies together publish 60% of that type of publication. "Publishing Scientific Journals Online," *BioScience, 47*(3), 1997, http://www.aibs.org/latitude/latpublications.html>. In 1997, the following distribution was shown by our data. Learned societies' journals represented 55% of journals and commercial publishers' journals, 44%. Boismenu, Gérard, and Guylaine Beaudry, "Publications

électroniques et revues savantes: Acteurs, rôles et réseaux," *Documentaliste. Sciences de l'information, 36* (6), November/December 1999, pp. 292–305, and *Documentation et Bibliothèques,45* (4), October-December 1999, pp. 149–59.

30   Our calculations are based on the cost of a one-year institutional subscription for a US institution, in US currency. The cost of a subscription to the print version is used as a reference, where the publisher allows a choice of subscribing to the print version, the electronic version or both.

31   The data are comparable to the results published by *Library Journal* in April 1999.

32   Note that the most expensive journals in our sample published by a learned society cost nearly US$6,020 per year (*A Mathematical and General Journal of Physics*, American Institute of Physics), while the most expensive of the commercial journals cost slightly more than US$8,630 (*Physics Letters B*, published by North-Holland/Elsevier).

33   Again, commercial publishers have the distinction of the most expensive journal (US$1,310 for the journal *Child Abuse & Neglect*, also produced by Elsevier), followed by learned societies (US$365 for *American Political Science Review*, American Political Science Association) and university presses (US$295 for *British Journal of Social Work* and *Health Education Research* from Oxford University Press).

34   Brueggeman, Peter, *Cost per Page for 1996 Subscriptions Costing over $500 at SIO Library*, 1996, Scripps Institution of Oceanography Library Web site, <http://www.scilib.ucsd.edu/sio/guide/prices/perpage.html>.

35   Wilder, Stanley J., "Comparing Value and Estimated Revenue of SciTech Journals," *ARL Newsletter of Research Library Issues and Actions*, No. 200, 1998, <http://www.arl.org/newsltr/200/wilder.html>.

36   Similarly see Adler, Kraig, and Wally Olsen, "Journals Pricing," *Learned Publishing, 12* (2), 1999, pp. 137–39, and "University-Press Group to Study Whether Books in Some Fields Are Disappearing: Non-Profit Journals Are Found to Be More Cost-Effective than Commercial Ones," *The Chronicle of Higher Education,46* (6), 1999, <http://chronicle.com/weekly/v46/i06/06a02401.htm>.

37   The discussion in this chapter is based on a review of basic data that is more fully developed for Canada, our participation in the university community, the many discussions we have had with journal editors and our involvement in journal publishing in recent years. The analysis is thus based on a compilation of data and the observations and experience derived from our presence in the community.

38   Counting all journals funded by public agencies (SSHRC, NSERC and Fonds FCAR) and journals inventoried in secondary databases, such as SCI, SSCI and AHCI.

39   Banque de données sur les revues savantes québécoises. Godin, B., E. Archambault and F. Vallières, "La production scientifique québécoise: Mesure basée sur la BRSQ" [Scientific activity in Quebec as measured by the BRSQ], *Argus, 29* (1), 2000, pp. 15–23.

40 Loans of services can have very high institutional visibility or, conversely, be in a grey area in budget terms. With the budget restrictions in university institutions, a number of acquisitions have been questioned, particularly gateways between institutions and journal organizations. The disappearance in recent years of the SSHRC grant to support the operation of learned societies has had an impact on support for their journals. The decrease in the number of grants awarded by the Fonds FCAR in the last competition and the generally lower grant ceiling, in an attempt to assist the weakest players, have reduced the resources available for many journals. All those factors have resulted in more pressure on journal editorial teams, although we cannot assume or suggest they were already well-supplied with resources, on the contrary. In comparison with France, the CNRS provides financial support for nearly 200 journals, mainly support through services or editorial assistance. Although not in the same order of magnitude, the subscription level is often similar.

41 Data, for 1998, was provided by the SSHRC for a sample of twenty-seven journals subsidized by the SSHRC with administrative offices in Quebec, and compiled by the authors.

42 The Canadian Association of Learned Journals recently estimated the contribution at between C$5,000 and C$10,000 per year, depending on the institution, <http://calj.icaap.org/>.

43 Godin, B., E. Archambault and F. Vallières, "La production scientifique québécoise: Mesure basée sur la BRSQ," *Argus, 29* (1), 2000, pp. 15–23; Godin, B., "Les revues savantes québécoises: Entre impact national et visibilité internationale," *Recherches sociographiques*, in press.

44 Data compiled from an SSHRC sample.

45 Boismenu, Gérard, and Guylaine Beaudry, *Pénétration des revues savantes canadiennes dans les universités américaines.* Mimeograph, 2001, 20 p.

CHAPTER 2

1 Odlyzko, Andrew, "Competition and Cooperation: Libraries and Publishers in the Transition to Electronic Scholarly Journals," *The Journal of Electronic Publishing, 4* (4), 1999, <http://www.press.umich.edu/jep/04-04/odlyzko0404.html>.

2 Mollier, Jean-Yves, et al., *Où va le livre?* Paris: La Dispute, 2000, 283 p.; "Édition, Éditeurs (1) et (2)," *Actes de la recherche en sciences sociales*, No. 126–27, March 1999, No. 130, December 1999; Schiffrin, André, *L'édition sans éditeurs?* Mayenne: La fabrique-Éditions, 1999, 94 p.

3 Vettraino-Soulard, Marie-Claude, *Les enjeux culturels d'Internet*, Paris: Hachette, 1998, p. 65, translation.

4 Fletcher, Lloyd Alan, "Developing an Integrated Approach to Electronic Publishing: Tailoring your Content for the Web," *Learned Publishing, 12* (2),

1999, pp. 107–17; Houghton, John W., "Crisis and Transition: The Economics of Scholarly Communication," *Learned Publishing, 14* (3), 2001, pp. 167–76; Caplan, Priscilla, *Report on the NISO/NFAIS, Workshop: Electronic Journals – Best Practices*, National Information Standards Organization (NISO), February 2000, <http://www.niso.org/e-jrnl-report.html>; Le Crosnier, Hervé, *Avons-nous besoin des journaux électroniques?* Paper presented at Journées SFIC-ENSSIB, *Une nouvelle donne pour les journaux scientifiques*, Villeurbanne, November 20, 1997, <http://www.info.unicaen.fr/herve/pub97/enssib/enssib.html>.

5   Raney, Keith R., "Into a Glass Darkly," *The Journal of Electronic Publishing, 4* (2), December 1998, <http://www.press.umich.edu/jep/04-02/raney.html>.

6   Varian, Hal R., "The Future of Electronic Journals," *The Journal of Electronic Publishing, 4* (1), September 1998, <http://www.press.umich.edu/jep/04-01/varian.html>.

7   Tenopir, Carol, and Donald W. King, "Designing Electronic Journals with 30 Years of Lessons from Print," *The Journal of Electronic Publishing, 4* (2), December 1998, <http://www.press.umich.edu/jep/04-02/king.html>.

8   Abate, Tom, "Publishing Scientific Journals Online," *BioScience, 47* (3), 1997, <http://www.aibs.org/latitude/latpublications.html>.

9   Whisler, Sandra, and Susan F. Rosenblatt, *The Library and the University Press: Two Views of the Costs and Problems of the Current System of Scholarly Publishing*, Paper presented at the *Scholarly Communication Technology* conference, Emory University, April 1997, <http://www.arl.org/scomm/scat/rosenblatt.html>.

10   Odlyzko, Andrew, "The Economics of Electronic Journals," *The Journal of Electronic Publishing, 4* (1), September 1998, <http://www.press.umich.edu/jep/04-01/odlyzko.html>.

11   Varian, Hal R., "The Future of Electronic Journals," *The Journal of Electronic Publishing, 4* (1), September 1998, <http://www.press.umich.edu/jep/04-01/varian.html>.

12   Odlyzko, Andrew, "The Economics of Electronic Journals," *The Journal of Electronic Publishing, 4* (1), September 1998, <http://www.press.umich.edu/jep/04-01/odlyzko.html>; also "Competition and Cooperation: Publishers in the Transition to Electronic Scholarly Journals," *The Journal of Electronic Publishing, 4* (4), 1999, <http://www.press.umich.edu/jep/04-04/odlyzko0404.html>.

13   Tenopir, Carol, and Donald W. King, "Designing Electronic Journals with 30 Years of Lessons from Print," *The Journal of Electronic Publishing, 4* (2), December 1998, <http://www.press.umich.edu/jep/04-02/king.html>.

14   On the archiving issue, see, in particular, Le Crosnier, Hervé, *Avons-nous besoin des journaux électroniques?* Paper presented at Journées SFIC-ENSSIB, *Une nouvelle donne pour les journaux scientifiques*, Villeurbanne, November 20, 1997, <http://www.info.unicaen.fr/herve/pub97/enssib/enssib.html>.

15   See the discussion initiated by Beebe, Linda, and Barbara Meyers, "The Unsettled State of Archiving," *The Journal of Electronic Publishing, 4* (4), June 1999, <http://www.press.umich.edu/jep/04-04/beebe.html>; Rohe, Terry Ann, "How Does Electronic Publishing Affect the Scholarly

Communication Process?" *The Journal of Electronic Publishing, 3* (3), 1998, <http://www.press.umich.edu/jep/03-03/rohe.html>; Phillips, Margaret E., "Ensuring Long-Term Access to Online Publications," *The Journal of Electronic Publishing, 4* (4), June 1999, <http://www.press.umich.edu/jep/04-04/phillips.html>.

16 Robnett, Bill, *Online Journal Pricing*, The Haworth Press Inc., 1997, <http://web.mit.edu/waynej/www/robnett.htm>.

17 King, Donald W., and Carol Tenopir, *Economic Cost Models of Scientific Scholarly Journals*, Paper presented at the *ICSU Press Workshop*, Oxford, United Kingdom, March 1998, <http://www.bodley.ox.ac.uk/icsu/kingppr.html>, "Evolving Journal Costs: Implications for Publishers, Libraries, and Readers," *Learned Publishing, 12* (4), 1999, pp. 251–58.

18 Abate, Tom, "Publishing Scientific Journals Online," *BioScience, 47* (3), 1997, <http://www.aibs.org/latitude/latpublications.html>.

19 Whisler, Sandra, and Susan F. Rosenblatt, *The Library and the University Press: Two Views of the Costs and Problems of the Current System of Scholarly Publishing*, Paper presented at the *Scholarly Communication and Technology* conference, Emory University, April 1997, <http://www.arl.org/scomm/scat/rosenblatt.html>.

CHAPTER 3

1 "A New Approach to Finding Research Materials on the Web," CLIR Issues, No. 16, 2000, <http://www.clir.org/pubs/issues16.html#approach>.

2 See Lynch, Patrick J., and Sarah Horton, Yale Style Manual – Site Design, 1997, <http://info.med.yale.edu/caim/manual/contents.html>; Communications Jean Lalonde, Les bâtisseurs de l'inforoute – Site indépendant d'assistance et de référence en conception de site Web au Québec (Section: 5 questions pour éviter l'échec d'un site Web d'entreprise, 1996–98), <http://www.cjl.qc.ca/batisseurs/5questions.htm/>; Nielsen, Jakob, Top Ten Mistakes in Web Design, 1996, <http://www.useit.com/alertbox/9605.html/>; Nielsen Jakob, Top Ten New Mistakes in Web Design, 1999, <http://www.useit.com/alertbox/990530.html>.

3 Metadata are information on digital objects, such as a journal article. In that case, metadata describe the article's content. They are useful for retrieval, but also for administration, description, access and preservation of information. They identify, in particular, the article's content, subject, title, author's name, abstract and structure.

4 See <http://www.nstein.com> or <http://www.nominotechnologies.com>.

5 Tracey, Stanley, "Moving Up The Ranks," Ariadne, No. 12, November 1997, <http://www.ariadne.ac.uk/issue12/search-engines/>, "Keyword Spamming: Cheat Your Way To The Top," Ariadne, No. 10, July 1997,

<http://www.ariadne.ac.uk/issue10/search-engines/>; Bradley, Phil, "The relevance of underpants to searching the Web," Ariadne, No. 24, July 1997, <http://www.ariadne.ac.uk/issue24/search-engines/>; Liberatore, Karen, "Getting to the Source," Macworld, September 22, 2000, <http://macworld.zetnet.com/features/pov.4.4.html>; Murphy, Kathleen, "Cheaters Never Win," Internetworld, May 1996, <http://www.internetworld.com/print/1996/05/20/undercon/cheaters.html>.

6   Georgia Institute of Technology, Graphic, "Visualization and Usability Center," Tenth User Survey, 1998, <http://www.gvu.gatech.edu/user_surveys/survey-1998-10/>.

7   Hartley, Jakob, "Is it appropriate to use structured abstracts in social science journals?" Learned Publishing, 10(4), October 1997, pp. 313–17; See also the ISO/IEC standard, ISP 12059-13: 1996 on the production of abstracts.

8   <http://www.publiclibraryofscience.org/>; See the debate that initiative generated in Nature, <http://www.nature.com/nature/debates/e-access>; Ann Okerson concluded her article: "Extremist language and extremist imagery are out of place and have an obstructionist effect. If we can set aside extremism, I believe we can already see around us the elements of new forms of publication that are inspiring and encouraging." <http://www.nature.com/nature/debates/e-access/articles/okerson.html>.

9   Budapest Open Access Initiative, February 14, 2002, <http://www.soros.org/>.

10 Rowland, Fytton, "Electronic Publishing: Non-Commercial Alternatives," Learned Publishing, 12(3), 1999, pp. 209–16.

11 Keller, Michael, "Innovation and Service in Scientific Publishing Requires More, Not Less, Competition," Nature, <http://www.nature.com/nature/debates/e-access/Articles/keller.html>. Keller is the editor of HighWire Press.

12 This is based, as are the following estimates, on a sample of twenty-seven journals supported financially by the SSHRC. The sample was chosen on a random basis by the granting agency and represents half the funded journals with administrative offices (address of head office) in Quebec.

13 Two organizations contribute through competitions, but not all journals are in a position to obtain two grants (in total, thirty-three journals are subsidized by the Fonds FCAR and approximately fifty-five by the SSHRC). In France, the value of loans of services provided to journals by the CNRS must be counted in addition to annual financial grants.

14 Kiernan, Vincent, "Why Do Some Electronic-Only Journals Struggle, While Others Flourish?" The Chronicle of Higher Education, May 21, 1999, <http://chronicle.merit.edu/weekly/v45/i37a02501.htm>. This article was reproduced in The Journal of Electronic Publishing, 4(4), 1999, <http://www.press.umich.edu/jep/04-04/kiernan.html>; Lawrence, Steve, "Free Online Availability Substantially Increases a Paper's Impact,"

Nature, Webdebates, <http://www.nature.com/nature/debates/e-access/articles/lawrence.html>.

15 Slowinski, F. Hill, and Patrick Bernuth, "How 'Free Distribution' Impacts Your Business Model: Is It Really Free?" Learned Publishing, 14(2), 2001, pp. 144–48.

16 Butler, Declan, "Los Alamos loses physics archive as preprint pioneer heads east," Nature, <http://www.nature.com/nature/debates/e-access/articles/ginsparg.html>.

17 If questioning peer-review procedures and alternative proposals are utilized as administrative cost reduction mechanisms, we probably have to start by discussing the merits of the issue to see whether it is favourably received. Donavan, Bernard, "The Truth About Peer Review," Learned Publishing, 11(3), 1998, pp. 179–84; Nadasdy, Zoltan, "A Truly All-Electronic Journal: Let Democracy Replace Peer Review," The Journal of Electronic Publishing, 3(1), 1997, <http://www.press.umich.edu/jep/03-01/EJCBS.html>; Harnad, Stevan, "The Invisible Hand of Peer Review," Nature, November 5, 1998, <http://helix.nature.com/webmatters/invisible/invisible.html>; van Rooyen, Susan, "A Critical Examination of the Peer Process," Learned Publishing,11(3), 1998, pp. 185–91, and "The Evaluation of Peer-Review Quality," Learned Publishing,14(2), 2001, pp. 85–91.

18 HighWire Press, Bench>Press, <http://benchpress.highwire.org>; Beebe, Linda, and Barbara Meyers, "Digital Workflow: Managing the Process Electronically," The Journal of Electronic Publishing, 5(4), 2000, <http://www.press.umich.edu/jep/05-04/sheridan.html>; Wood, Dee, "Online Peer Review," Learned Publishing, 11(3), 1998, pp. 193–98.

19 For example, the New Journal of Physics, <http://njp.org/>.

20 This information was provided by Aldyth Holmes, Director, NRC Scientific Press. Implementation of the measure reveals certain points. First, the NRC financial contribution to the journals has never been reduced, and the way they are produced has not been changed by the measure. A new public player intervened in distribution, providing free access to Canadian users, and the price of that intervention was determined with reference to the journal's production cost, apart from the print media costs. Finally, filtered access for a fee was maintained for non-Canadian users and other organizations with an IP address without <.ca>, and that revenue is supposed to balance the whole NRC journal budget.

21 After three years of existence, the Project MUSE report pointed out, in 1998, that electronic dissemination had not caused the number of subscriptions to drop. At Blackwell's, it was noted that there was interest in electronic formats, but not a tendency to substitute electronic media for print media; instead, users prefer to combine the electronic and print version of the journal. However, it is nonetheless thought that subscriptions to the print version are slated to slowly disappear. In both cases, online access is not free. For MUSE, offering mainly institutional

32   Also, Johns Hopkins University Press, which proposed in 1998 that
     university presses join Project MUSE to benefit from the journal
     production and distribution services, has seen the number of journals
     participating in the project grow from forty-four to more than one
     hundred. The sale of subscription packages and subscriptions to
     the sectoral collection is becoming particularly attractive, <http://
     muse.jhu.edu/proj_descrip/subscribe.html>.

33   We did not consider the price level based on the quality of the electronic
     formats disseminated or the services offered.

34   We doubt there is a high demand for individual online subscriptions.
     In the near future, individual subscriptions should focus on the print
     version for individuals who really do not have access to the electronic
     version through libraries, particularly those that are resistant. It would
     then be appropriate to track reader reaction and make adjustments.

CHAPTER 4

1   Fletcher, Lloyd Alan, "Developing an Integrated Approach to Electronic
     Publishing: Tailoring your Content for the Web," Learned Publishing,
     12(2), 1999, pp. 107–17; Bachrach, Steven M., Darin C. Burleigh and
     Anatoli Krassivine, "Designing the Next-Generation Chemistry Journal:
     The Internet Journal of Chemistry," Issues in Science and Technology
     Librarianship, Winter 1998, <http://www.library.ucsb.edu/istl/98-winter/
     article1.html>.

2   Fletcher, Lloyd Alan, "Developing an Integrated Approach to Electronic
     Publishing: Tailoring your Content for the Web," Learned Publishing,
     12(2), 1999, pp. 107–17.

3   See also Clément, Caroline, and Marc Bonvin (edited by Françoise
     Khenoune), Les périodiques en sciences humaines et sociales, March
     2000, Bibliothèque cantonale et universitaire de Lausanne-Dorigny,
     <http://www.unil.ch/BCU/recherches/l_art_bi.htm>.

4   This chapter deals specifically with current journal production. The
     techniques, formats and procedures for retrospective digitization of
     documents for which computer files no longer exist is something entirely
     different.

5   Ockerbloom, John Mark, "Archiving and Preserving PDF files," RLG
     DigiNews, 5(1), February 2001, <http://www.rlg.org/preserv/diginews/
     diginews5-1.html#feature2>.

6   On this subject, see The Adobe XML Architecture, < http://
     www.adobe.com/enterprise/xml.html>.

7   Evidently, here again, variants may be introduced based on a set of factors
     specific to the production environment. The integrated XML chain can
     be adapted for particular conditions.

subscriptions that give access to individuals associated with institutions, consultation is de facto free for individuals. However, maintaining a subscription to the print version is primarily the responsibility of libraries. In the second case, access is filtered, and individuals and institutions pay a fee. Although we may select those examples for discussion on the extent and limitations of filtered access for a fee (which follows), the observations made here cannot be transferred unreservedly, because free access on the Web changes the overall environment for users. Marie R. Hansen, Dir., Project MUSE, Process Report, The Johns Hopkins University Press with The Milton S. Eisenhower Library, January 1995 to June 1998, 40 p. Hodson, Richard, "The demand for journals – fact versus fiction," Learned Publishing, 11(3), July 1998, pp. 205–8.

22   <http://muse.jhu.edu/>.

23   <http://highwire.stanford.edu/>.

24   <http://www.bioone.org/bioone/?request=index-html>.

25   Day, Colin, Pricing Electronic Products, Paper presented at AAUP/ARL Symposium on Electronic Publishing, November 1994, <http://www.press.umich.edu/jep/works/colin.eprice.html>.

26   Varian, Hal R., "Pricing Electronic Journals," D-Lib Magazine, June 1996, <http://www.dlib.org/dlib/june96/06varian.html>.

27   See University of Chicago Press for journals such as The American Journal of Human Genetics, The Astronomical Journal, Current Anthropology and Astronomical Journal. They are essentially journals published for learned societies, <http://www.journals.uchicago.edu>. The publications of organizations such as the Association for Computing Machinery, Institute of Physics, American Institute of Physics and Australian Academy of Science could also be cited. Robnett, Bill, Online Journal Pricing, The Haworth Press Inc., 1997, <http://web.mit.edu/waynej/www/robnett.htm>.

28   A number of journals that joined HighWire Press practise that type of pricing, to name a few, AJP: Endocrinology and Metabolism, Genetics, Journal of Applied Physiology and Physiological Reviews, <http://highwire.stanford.edu/>.

29   We are referring here to journals such as Evolutionary Computation, Journal of Economics & Management Strategy, The Quarterly Journal of Economics and The Journal of Interdisciplinary History, <http://mitpress.mit.edu/>.

30   <http://highwire.stanford.edu/institutions>. According to the assistant editor of the journal, Robert Simoni, the price of the electronic version was set to enhance the autonomy and independence of the electronic journal in relation to the print version so that it does not appear to be an adjunct to the print version. Robnett, Bill, Online Journal Pricing, The Haworth Press Inc., 1997, <http://web.mit.edu/waynej/www/robnett.htm>.

31   <http://www.press.jhu.edu/press/journals/>.

CHAPTER 5

1   Raney, R. Keith, "Into a Glass Darkly," The Journal of Electronic
    Publishing, 4(2), December 1998, <http://www.press.umich.edu/jep/04-
    02/raney.html>.

2   On the expectations authors and readers have of journals, see also Morris,
    Sally, "Learned Journals and the Communication of Research," Learned
    Publishing, 11(4), 1998, pp. 253–58, and "What Authors Want: The ALPSP
    Research Study on the Motivations and Concerns of Contributors to
    Learned Journals," Learned Publishing, 12(3), 1999, pp. 170–72.

3   Donovan, Bernard, "The Truth About Peer Review," Learned Publishing,
    11(3), 1998, pp. 179–84; Nadasdy, Zoltan, "A Truly All-Electronic Journal:
    Let Democracy Replace Peer Review," The Journal of Electronic Publishing,
    3(1), 1997, <http://www.press.umich.edu/jep/03-01/EJCBS.html>;
    Harnard, Stevan, "The Invisible Hand of Peer Review," Nature, November
    5, 1998, <http://helix.nature.com/nature/webmatters/invisible/
    invisible.html>; van Rooyen, Susan, "A Critical Examination of the Peer
    Process," Learned Publishing, 11(3), 1998, pp. 185–91, and "The Evaluation
    of Peer-Review Quality," Learned Publishing,14(2), 2001, pp. 85–91.

4   See, in particular, the services at HighWire, HighWire Press, Bench>Press,
    <http://benchpress.highwire.org>.

5   Beebe, Linda, and Barbara Meyers, "Digital Workflow: Managing the
    Process Electronically," The Journal of Electronic Publishing, 5(4), 2000,
    <http://www.press.umich.edu/jep/05-04/sheridan.html>.

6   On this subject, see the comments of Marc Guillaume, L'empire des
    réseaux, Paris: Descartes & Cie, 1999, 158 p., translation; See also Fisher,
    Hervé, Le Choc du numérique, Montreal, VLB Éditeur, 2001, pp. 19–88.

7   Le Crosnier, Hervé, Avons-nous besoin des journaux électroniques?
    Paper presented at Journées SFIC-ENSSIB, Une nouvelle donne pour
    les journaux scientifiques, Villeurbanne, November 20, 1997, <http:
    //www.info.unicaen.fr/herve/pub97/enssib/enssib.html>.

8   Note here that review reports do not necessarily act as a counterbalance;
    although specialists read candidates' publications, they are influenced by
    the same indicators themselves and often attach a great deal of importance
    to them.

9   Memorandum, 1999–2000 Academic Reappointment/Promotion
    Instructions, Rutgers, The State University of New Jersey, June 10, 1999,
    <http://www.rutgers.edu/oldqueens/instruct.doc>.

10 Kiernan, Vincent, "Why Do Some Electronic-Only Journals Struggle,
    While Others Flourish?" The Chronicle of Higher Education, May 21, 1999,
    <http://chronicle.merit.edu/weekly/v45/i37/37a02501.htm>. This article
    was reprinted in The Journal of Electronic Publishing,4 (4), 1999, <http:
    //www.press.umich.edu/jep/04-04/kiernan.html>.

11   Sweeney, Aldrin E., "Should You Publish in Electronic Journals?" The Journal of Electronic Publishing, 6(2), December 2000, <http://www.press.umich.edu/jep/06-02/sweeney.html>.

12   Wyly, Brendan J., Competition in Scholarly Publishing? What Publisher Profits Reveal, <http://www.arl.org/newsltr/200/wyly.html>.

13   Langston, Lizbeth, Scholarly Communication and Electronic Publication: Implications for Research, Advancement, and Promotion, Untangling the Web, Santa Barbara Library Web page, University of California, <http://www.library.zucsb.edu/untangle/langston.html>.

14   Fosmire, Michael, and Song Yu, "Free Scholarly Electronic Journals: How Good Are They?" Issues in Science and Technology Librarianship, Summer 2000, <http://www.library.ucsb.edu/istl/00-summer/refereed.html>.

15   Kiernan, Vincent, "Why Do Some Electronic-Only Journals Struggle, While Others Flourish?" The Chronicle of Higher Education, May 21, 1999, <http://chronicle.merit.edu/weekly/v45/i37/37a02501.htm>. This article was reprinted in The Journal of Electronic Publishing, 4(4), 1999, <http://www.press.umich.edu/jep/04-04/kiernan.html>.

16   Varian, Hal R., "The Future of Electronic Journals," The Journal of Electronic Publishing, 4(1), September 1998, <http://www.press.umich.edu/jep/04-01/varian.html>.

17   Abate, Tom, "Publishing Scientific Journals Online," BioScience, 47(3), 1997, <http://www.aibs.org/latitude/latpublications.html>.

18   Whisler, Sandra, and Susan F. Rosenblatt, The Library and the University Press: Two Views of the Costs and Problems of the Current System of Scholarly Publishing, Paper presented at the Scholarly Communication Technology conference, Emory University, April 1997, <http://www.arl.org/scomm/scat/rosenblatt.html>.

19   Guillaume, Marc (Ed.), Où vont les autoroutes de l'information? Paris: Descartes & Cie, 1997, 190 p.

20   Ibid., p. 96, translation, emphasis added.

21   Kling, Rob, and Lisa Covi, "Electronic Journals and Legitimate Media in the Systems of Scholarly Communication," The Information Society, 11(4), 1995, pp. 261–71, <http://www.ics.uci.edu/~kling/klingej2.html>.

22   Kiernan, Vincent, "Why Do Some Electronic-Only Journals Struggle, While Others Flourish?" The Chronicle of Higher Education, May 21, 1999, <http://chronicle.merit.edu/weekly/v45/i37/37a02501.htm>. This article was reprinted in The Journal of Electronic Publishing, 4(4), 1999, <http://www.press.umich.edu/jep/04-04/kiernan.html>.

23   See, in particular, these books: Castells, Manuel, La société en réseaux. L'ère de l'information, Paris: Fayard, 1998, 613 p., and La galaxie Internet, Paris: Fayard, 2001, 365 p.; Guillaume, Marc, L'empire des réseaux, Paris: Descartes & Cie, 1999, 158 p.; Berners-Lee, Tim, Weaving the Web, New York: Harper Collins, 2000, 246 p.; Brown, John Seely, and Paul Duguid, The Social Life of Information, Boston: Harvard Business School Press, 2000, 320 p.; Carnoy, Martin, Dans quel monde vivons-nous? Paris: Fayard,

2001, 351 p.; Arms, William, Digital Libraries, MIT Press; Commissariat général du Plan, Les réseaux de la société de l'information, Paris: Éditions ESKA, Official Reports Collection, 1996, 224 p.; Fdida, Serge, Des autoroutes de l'information au cyberespace, Paris: Flammarion, Dominos, 1997, 123 p.; Collective, La radio à l'ère de la convergence, Montreal: Les Presses de l'Université de Montréal, 2001, 205 p.; Wolton, Dominique (with Olivier Jay), Internet: Petit manuel de survie, Paris: Flammarion, 2000, 186 p.; de Kerckhove, Derrick, L'intelligence des réseaux, Paris: Éditions Odile Jacob, 2000, 303 p.; Sicard, Marie-Noële, and Jean-Michel Besnier (Eds.), Les Technologies de l'information et de la communication: Pour quelle société? L'Université de technologie de Compiègne, 1998, 271 p.

24 Castells, Manuel, La société en réseaux. L'ère de l'information, Paris: Fayard, 1998, p. 525, translation.
25 Ibid., p. 87 et seq.
26 Fletcher, Lloyd Alan, "Developing an Integrated Approach to Electronic Publishing: Tailoring your Content for the Web," Learned Publishing, 12(2), 1999, pp. 107–17.

498100